计算机视觉

核心技术、算法与开发实战

王 丽◎编著

中国铁道出版社有限公司
CHINA RAILWAY PUBLISHING HOUSE CO., LTD.

北 京

内 容 简 介

本书详细讲解了使用 Python 语言开发 AI 图像视觉处理程序的知识。全书共 11 章，依次讲解了图像视觉技术基础，图像的采样、变换和卷积处理，图像增强处理，图像特征提取处理，图像分割处理，目标检测处理，图像分类处理，鲜花识别系统开发，智能素描绘图系统开发，小区 AI 停车计费管理系统开发和机器人智能物体识别系统开发。书中在详细讲解每个知识点的同时，还穿插了大量的实例来演示每个知识点的用法，引领读者扎实掌握基于 AI 的图像视觉开发技术。

本书适用于已经了解了 Python 语言基础语法，想进一步学习计算机视觉开发的读者，还可以作为大专院校计算机科学及相关专业的师生用书和培训学校的参考书。

图书在版编目（CIP）数据

计算机视觉：核心技术、算法与开发实战/王丽编著. —北京：
中国铁道出版社有限公司，2024.6
ISBN 978-7-113-31152-0

Ⅰ.①计… Ⅱ.①王… Ⅲ.①计算机视觉 Ⅳ.①TP302.7

中国国家版本馆 CIP 数据核字(2024)第 073234 号

书　　名：计算机视觉——核心技术、算法与开发实战
JISUANJI SHIJUE：HEXIN JISHU SUANFA YU KAIFA SHIZHAN

作　　者：王　丽

责任编辑：于先军　　编辑部电话：（010）51873026　　电子邮箱：46768089@qq.com
封面设计：宿　萌
责任校对：苗　丹
责任印制：赵星辰

出版发行：中国铁道出版社有限公司（100054，北京市西城区右安门西街 8 号）
网　　址：http://www.tdpress.com
印　　刷：天津嘉恒印务有限公司
版　　次：2024 年 6 月第 1 版　　2024 年 6 月第 1 次印刷
开　　本：787 mm×1 092 mm　1/16　印张：17.5　字数：426 千
书　　号：ISBN 978-7-113-31152-0
定　　价：79.80 元

版权所有　侵权必究

凡购买铁道版图书，如有印制质量问题，请与本社读者服务部联系调换。电话：（010）51873174
打击盗版举报电话：（010）63549461

配套资源下载网址：
http://www.m.crphdm.com/2024/0326/14703.shtml

前　言

随着计算能力的提升和深度学习技术的兴起，图像视觉领域已经取得了令人瞩目的进展。从卫星图像分析到医学图像诊断，图像视觉的应用领域正在不断拓展。然而，学习该技术并将其很好地应用于实践，需要深入的知识和实际经验。

本书是一本较为全面地介绍图像视觉开发和实际应用的书籍，内容涵盖了广泛的主题，包括图像识别、采样、变换、卷积处理、增强处理、特征提取、分割处理、目标检测及分类处理，同时讲解了多个实际项目案例。具体内容如下：

第 1 章介绍了图像视觉的基本概念，包括图像识别的发展、应用和基本步骤，以及与 AI、机器学习和深度学习相关的核心概念。

第 2 章介绍了图像采样、离散傅里叶变换、卷积处理和频域滤波等技术。

第 3 章介绍了如何提高图像质量和增强特定特征，包括对比度增强、锐化、噪声减少、色彩平衡、超分辨率和去除运动模糊等技术。

第 4 章讲解了如何从图像中提取有用的信息，包括颜色、纹理、形状和斑点等特征。

第 5 章介绍了图像分割的不同方法，如基于阈值、边缘、区域、图论和深度学习的分割技术。

第 6 章介绍了目标检测的步骤、方法和实际应用，包括 YOLO、语义分割、SSD 目标检测。

第 7 章探讨了图像分类的不同方法，包括基于特征提取和机器学习、卷积神经网络、迁移学习、循环神经网络和卷积循环神经网络等。

第 8~11 章是实战项目案例，详细介绍了鲜花识别系统、智能素描绘图系统、小区 AI 停车计费管理系统和机器人智能物体识别系统的开发流程与具体实现方法。

书中通过理论和实际案例的结合来讲解图像视觉开发知识，使读者能够掌握从图像处理到实际应用的关键技能，以满足日益增长的图像视觉领域的需求。本书主要特色如下：

内容全面：书中涵盖了图像视觉领域的主要内容，从基础知识到高级应用，使读者能够系统学习图像视觉技术。

讲解细致：每个案例都提供了详细的步骤和代码，带领读者逐步实施和理解项目。

理论与实践结合：除了必要的理论概念，书中还提供了多个实际项目案例，帮助读者将所学知识应用到实际问题中，从而更好地掌握所学技能。

项目驱动学习：每一章的项目案例都以实际问题为基础，读者通过构建和开发项目来学习相关概念和技术。

实用性强：书中讲解的技术和方法都是实际应用中经常使用的，读者可以将所学知识直接应用到自己的项目和工作中。

本书的内容适用于各种不同背景和经验水平的读者，从初学者到专业人士，都可以从中受益。无论是学习图像视觉的基础知识还是应用这些知识到实际项目中，本书都有讲解。具体适用于以下读者：

- 计算机科学及其相关专业的师生；
- 软件工程师和开发人员；
- 数据科学家和研究人员；
- 人工智能研究者；
- 图像处理爱好者。

本书在编写过程中，得到了中国铁道出版社有限公司编辑们的大力支持，正是各位编辑的求实、耐心和效率，才使本书能够在这么短的时间内出版。另外，也十分感谢我的家人给予的巨大支持。书中存在的纰漏或不足之处诚请读者指正。

王 丽

2024 年 4 月

目　录

第1章　图像视觉技术基础

1.1　图像识别概述 1
   1.1.1　什么是图像识别 1
   1.1.2　图像识别的发展阶段 1
   1.1.3　图像识别的应用 2
1.2　图像识别的步骤 2
1.3　图像识别技术 3
   1.3.1　人工智能 3
   1.3.2　机器学习 4
   1.3.3　深度学习 4
   1.3.4　基于神经网络的图像识别 4
   1.3.5　基于非线性降维的图像识别 5

第2章　图像的采样、变换和卷积处理

2.1　采样 6
   2.1.1　最近邻插值采样 6
   2.1.2　双线性插值 8
   2.1.3　双立方插值 11
   2.1.4　lanczos 插值 13
2.2　离散傅里叶变换（DFT） 14
   2.2.1　为什么使用 DFT 14
   2.2.2　使用 NumPy 库实现 DFT 15
   2.2.3　使用 SciPy 库实现 DFT 16
   2.2.4　用快速傅里叶变换算法计算 DFT 17
2.3　卷积 18
   2.3.1　为什么需要卷积图像 18
   2.3.2　使用 SciPy 库中的函数 convolve2d()进行卷积操作 19
   2.3.3　使用 SciPy 库中的函数 ndimage.convolve()进行卷积操作 20
2.4　频域滤波 22
   2.4.1　什么是滤波器 22
   2.4.2　高通滤波器 22
   2.4.3　低通滤波器 23
   2.4.4　DoG 带通滤波器 24
   2.4.5　带阻滤波器 26

第3章 图像增强处理

- 3.1 对比度增强 ... 28
  - 3.1.1 直方图均衡化 ... 28
  - 3.1.2 自适应直方图均衡化 ... 30
  - 3.1.3 对比度拉伸 ... 32
  - 3.1.4 非线性对比度增强 ... 34
- 3.2 锐化 ... 35
  - 3.2.1 锐化滤波器 ... 36
  - 3.2.2 高频强调滤波 ... 39
  - 3.2.3 基于梯度的锐化 ... 42
- 3.3 噪声减少 ... 46
  - 3.3.1 均值滤波器 ... 46
  - 3.3.2 中值滤波器 ... 47
  - 3.3.3 高斯滤波器 ... 48
  - 3.3.4 双边滤波器 ... 49
  - 3.3.5 小波降噪 ... 49
- 3.4 色彩平衡 ... 50
  - 3.4.1 白平衡 ... 51
  - 3.4.2 颜色校正 ... 52
  - 3.4.3 调整色调和饱和度 ... 53
- 3.5 超分辨率 ... 54
- 3.6 去除运动模糊 ... 56
  - 3.6.1 边缘 ... 56
  - 3.6.2 逆滤波 ... 57
  - 3.6.3 统计方法 ... 58
  - 3.6.4 盲去卷积 ... 59

第4章 图像特征提取处理

- 4.1 图像特征提取方法 ... 61
- 4.2 颜色特征 ... 61
  - 4.2.1 颜色直方图 ... 62
  - 4.2.2 其他颜色特征提取方法 ... 63
- 4.3 纹理特征 ... 65
  - 4.3.1 灰度共生矩阵 ... 65
  - 4.3.2 方向梯度直方图 ... 66
  - 4.3.3 尺度不变特征变换 ... 67
  - 4.3.4 小波变换 ... 68
- 4.4 形状特征 ... 70
  - 4.4.1 边界描述子 ... 70
  - 4.4.2 预处理后的轮廓特征 ... 73
  - 4.4.3 模型拟合方法 ... 75

    4.4.4 形状上的变换 78
  4.5 基于 LoG、DoG 和 DoH 的斑点检测器 80
    4.5.1 LoG 80
    4.5.2 DoG 81
    4.5.3 DoH 82

第 5 章 图像分割处理

  5.1 图像分割的重要性 84
  5.2 基于阈值的分割 84
    5.2.1 灰度阈值分割 85
    5.2.2 彩色阈值分割 85
  5.3 基于边缘的分割 86
    5.3.1 canny 边缘检测 86
    5.3.2 边缘连接方法 87
  5.4 基于区域的分割 88
    5.4.1 区域生长算法 89
    5.4.2 基于图论的分割算法 90
    5.4.3 基于聚类的分割算法 92
  5.5 最小生成树算法 94
  5.6 基于深度学习的分割 95
    5.6.1 FCN（全卷积网络）.. 96
    5.6.2 U-Net 97
    5.6.3 DeepLab 97
    5.6.4 Mask R-CNN 98

第 6 章 目标检测处理

  6.1 目标检测介绍 100
    6.1.1 目标检测的步骤 100
    6.1.2 目标检测的方法 100
  6.2 YOLO v5 100
    6.2.1 YOLO v5 的改进 101
    6.2.2 基于 YOLO v5 的训练、验证和预测 101
  6.3 语义分割 114
    6.3.1 什么是语义分割 114
    6.3.2 DeepLab 语义分割 116
  6.4 SSD 目标检测 118
    6.4.1 摄像头目标检测 118
    6.4.2 基于图像的目标检测 119

第 7 章 图像分类处理

  7.1 图像分类介绍 121

7.2 基于特征提取和机器学习的图像分类 121
    7.2.1 基本流程 122
    7.2.2 基于 scikit-learn 机器学习的图像分类 122
    7.2.3 分类算法 125
    7.2.4 聚类算法 127
7.3 基于卷积神经网络的图像分类 129
    7.3.1 卷积神经网络基本结构 129
    7.3.2 第一个卷积神经网络程序 132
    7.3.3 使用卷积神经网络进行图像分类 136
7.4 基于迁移学习的图像分类 146
    7.4.1 迁移学习介绍 146
    7.4.2 基于迁移学习的图片分类器 147
7.5 基于循环神经网络的图像分类 150
    7.5.1 循环神经网络介绍 150
    7.5.2 实战演练 151
7.6 基于卷积循环神经网络的图像分类 151
    7.6.1 卷积循环神经网络介绍 152
    7.6.2 CRNN 图像识别器 152

第 8 章 鲜花识别系统开发

8.1 系统介绍 155
8.2 创建模型 155
    8.2.1 创建 TensorFlow 数据模型 155
    8.2.2 将 Keras 模型转换为 TensorFlow Lite 160
    8.2.3 量化处理 161
    8.2.4 更改模型 162
8.3 识别器的具体实现 163
    8.3.1 准备工作 163
    8.3.2 页面布局 165
    8.3.3 实现 UI Activity 167
    8.3.4 实现主 Activity 168
    8.3.5 图像转换 172
    8.3.6 使用 GPU 委托加速 176

第 9 章 智能素描绘图系统开发

9.1 背景介绍 177
9.2 需求分析 177
9.3 功能模块 178
9.4 预处理 179
    9.4.1 低动态范围配置 179
    9.4.2 图像处理和调整 180
    9.4.3 获取原始图像的笔画 181

         9.4.4 方向检测 184
         9.4.5 去蓝处理 187
         9.4.6 图像合成 188
         9.4.7 快速排序 191
         9.4.8 侧窗滤波 192
    9.5 开始绘图 194
         9.5.1 基于边缘绘画的绘图程序 195
         9.5.2 绘制铅笔画 203

第 10 章 小区 AI 停车计费管理系统开发

    10.1 背景介绍 205
    10.2 系统功能分析和模块设计 205
         10.2.1 功能分析 205
         10.2.2 系统模块设计 206
    10.3 系统 GUI 206
         10.3.1 设置基本信息 206
         10.3.2 绘制操作按钮 207
         10.3.3 绘制背景和文字 207
    10.4 车牌识别和收费 208
         10.4.1 登记业主的车辆信息 208
         10.4.2 识别车牌 208
         10.4.3 计算停车时间 209
         10.4.4 识别车牌并计费 210
    10.5 主程序 213

第 11 章 机器人智能物体识别系统开发

    11.1 背景介绍 216
    11.2 物体识别 216
         11.2.1 物体识别介绍 217
         11.2.2 物体识别的挑战 217
         11.2.3 图像特征的提取方法 218
    11.3 系统介绍 219
    11.4 准备模型 220
         11.4.1 模型介绍 220
         11.4.2 自定义模型 221
    11.5 基于 Android 的机器人智能检测器 223
         11.5.1 准备工作 223
         11.5.2 页面布局 225
         11.5.3 实现主 Activity 228
         11.5.4 物体识别界面 233
         11.5.5 摄像机预览界面拼接 235
         11.5.6 lib_task_api 方案 242

11.5.7　lib_interpreter 方案 243
11.6　基于 iOS 的机器人智能检测器 247
　　11.6.1　系统介绍 247
　　11.6.2　视图文件 248
　　11.6.3　摄像机处理 259
　　11.6.4　处理 TensorFlow Lite 模型 264
11.7　调试运行 270

第 1 章　图像视觉技术基础

图像识别技术是一种利用计算机视觉技术对图像进行分析和理解的方法。图像识别的目标是使计算机能够自动识别和理解图像中的对象、场景和特征；它是深度学习算法的一种实践应用。本章详细讲解图像识别技术的基础知识，为大家步入本书后面知识的学习打下基础。

1.1　图像识别概述

当我们看到一个东西，大脑会迅速判断是不是见过这个东西或者类似的东西。这个过程有点儿像搜索，我们把看到的东西和记忆中相同或类似的东西进行匹配，从而识别它。用机器进行图像识别的原理也是类似的，通过分类并提取重要特征而排除多余的信息来识别图像。机器的图像识别和人类的图像识别原理相近，过程也大同小异，只是技术的进步让机器不但能像人类一样认花认草认物认人，还开始拥有超越人类的识别能力。

1.1.1　什么是图像识别

图像识别是人工智能的一个重要领域，是指利用计算机对图像进行处理、分析和理解，以识别各种不同模式的目标和对象的技术，并对质量不佳的图像进行一系列的增强与重建，从而有效改善图像质量。

虽然人类的识别能力很强大，但随着社会的高速发展，人类自身识别能力已经满足不了我们的需求，于是就产生了基于计算机的图像识别技术。这就像人类研究生物细胞，完全靠肉眼观察细胞是不现实的，于是就产生了显微镜等用于精确观测的仪器。通常一个领域有固有技术无法解决的需求时，就会产生相应的新技术。图像识别技术也是如此，此技术的产生就是为了让计算机代替人类去处理大量的物理信息，解决人类无法识别或者识别率特别低的信息。

随着计算机及信息技术的迅速发展，图像识别技术的应用逐渐扩大到诸多领域，尤其是在面部及指纹识别、卫星云图识别及临床医疗诊断等多个领域日益发挥着重要作用。在日常生活中，图像识别技术的应用也十分普遍，比如车牌捕捉、商品条码识别及手写文字识别等。随着该技术的逐渐发展并不断完善，未来将具有更加广泛的应用领域。

1.1.2　图像识别的发展阶段

图像识别的发展经历了三个阶段，分别是文字识别、数字图像处理与识别及物体识别，具体说明如下：

- 文字识别的研究是从 1950 年开始的，一般是识别字母、数字和符号，从印刷文字识别到手写文字识别，应用非常广泛。
- 数字图像处理和识别的研究开始于 1965 年。数字图像与模拟图像相比具有图像清晰，存储、传输方便,传输过程中不易失真,处理方便等巨大优势,这些都为图像识别技术的发展提供了强大的动力。
- 物体的识别主要指的是对三维世界的客体及环境的感知和认识,属于高级的计算机视觉范畴。它是以数字图像处理与识别为基础的结合人工智能、系统学等学科的研究方向,其研究成果被广泛应用在各种工业及探测机器人上。

1.1.3 图像识别的应用

移动互联网、智能手机以及社交网络的发展带来了海量图片信息,不受地域和语言限制的图片逐渐取代了文字,成为了传词达意的主要方式。但伴随着图片成为互联网中的主要信息载体,很多难题也随之出现。当信息由文字记载时,我们可以通过关键词搜索轻易找到所需内容并进行任意编辑。但是当信息是由图片记载时,我们无法对图片中的内容进行检索,从而影响了从图片中找到关键内容的效率。图片给我们带来了快捷的信息记录和分享方式,却降低了我们的信息检索效率。在这种情况下,计算机的图像识别技术就显得尤为重要。

(1)图像识别的初级应用

在现实应用中,图像识别的初级应用主要是娱乐化、工具化,在这个阶段用户主要是借助图像识别技术来满足某些娱乐化需求。例如,百度魔图的"大咖配"功能可以帮助用户找到与其长相最匹配的明星,百度的图片搜索可以找到相似的图片;Facebook 研发了根据相片进行人脸匹配的 DeepFace;国内专注于图像识别的创业公司旷视科技成立了 VisionHacker 游戏工作室,借助图形识别技术研发移动端的体感游戏。

在图像识别的初级应用中还有一个非常重要的细分领域——光学字符识别(optical character recognition,OCR),是指光学设备检查纸上打印的字符,通过检测暗、亮的模式确定其形状,然后用字符识别方法将形状翻译成计算机文字的过程,就是计算机对文字的阅读。借助 OCR 技术,可以将这些文字和信息提取出来。在这方面,国内的产品有百度的涂书笔记和百度翻译等;而谷歌借助经过 DistBelief 训练的大型分布式神经网络,对于 Google 街景图库的上千万门牌号的识别率超过 90%,每天可识别百万门牌号。

(2)图像识别的高级应用

图像识别的高级应用主要是指拥有视觉的机器,当机器真正具有了视觉之后,它们完全有可能代替我们去做很多事情。目前的图像识别应用就像是盲人的导盲犬,在盲人行动时为其指引方向;而未来的图像识别技术将会同其他人工智能技术融合在一起成为盲人的全职管家。

1.2 图像识别的步骤

概括来说,图像识别的过程主要包括以下四个步骤:

(1)获取信息:主要是指将声音和光等信息通过传感器向电信号转换,也就是对识别对象的基本信息进行获取,并将其向计算机可识别的信息转换。

(2)信息预处理:主要是指采用去噪、变换及平滑等操作对图像进行处理,基于此使图像的重要特点提高。

(3)抽取及选择特征:主要是指在模式识别中,抽取及选择图像特征,概括而言就是识别图像具有种类多样的特点，如采用一定方式分离，就要识别图像的特征，获取特征也被称为特征抽取；在特征抽取中所得到的特征也许对此次识别并不都是有用的，这个时候就要提取有用的特征，这就是特征的选择。特征抽取和选择在图像识别过程中是非常关键的技术之一，所以对这一步的理解是图像识别的重点。

（4）设计分类器及分类决策：其中设计分类器就是根据训练对识别规则进行制定，基于此识别规则能够得到特征的主要种类，进而使图像识别的辨识率不断提高，此后再通过识别特殊特征，最终实现对图像的评价和确认。

在使用计算机进行图像识别的应用中，计算机首先就能够完成图像分类并选出重要信息，排除冗余信息，根据这一分类计算机就能够结合自身记忆存储结合相关要求进行图像的识别，这一过程本身与人脑识别图像并不存在着本质差别。对于图像识别技术来说，其本身提取出的图像特征直接关系着图像识别能否取得较为满意的结果。

值得注意的是，归根结底，毕竟计算机不同于人类的大脑，所以计算机提取出来的图像特征存在着不稳定性，这种不稳定性往往会影响图像识别的效率与准确率。这种情况下，在图像识别中引入 AI 技术就变得十分重要了。

1.3 图像识别技术

计算机的图像识别技术就是模拟人类的图像识别过程，在图像识别的过程中进行模式识别是必不可少的。本节详细讲解现实中主流的图像识别技术。

1.3.1 人工智能

人工智能就是我们平常所说的 AI，全称是 Artificial Intelligence。人工智能是研究、开发用于模拟、延伸和扩展人类智能的理论、方法、技术及应用系统的一门新的技术科学。人工智能由不同的领域组成，如机器学习、计算机视觉等，总的说来，人工智能研究的一个主要目标是使机器能够胜任一些通常需要人类智能才能完成的复杂工作。

人工智能单从字面上理解，应该理解为人类创造的智能。那么什么是智能呢？如果人类创造了一个机器人，这个机器人能有像人类一样甚至超过人类的推理、知识、学习、感知处理等这些能力，那么就可以将这个机器人称为是一个有智能的物体，也就是人工智能。

现在通常将人工智能分为弱人工智能和强人工智能，我们看到电影里的一些人工智能大部分都是强人工智能，它们能像人类一样思考如何处理问题，甚至能在一定程度上做出比人类更好的决定，它们能自适应周围的环境，解决一些程序中没有遇到的突发事件，具备这些能力的就是强人工智能。但是在目前的现实世界中，大部分人工智能只是实现了弱人工智能，能够让机器具备观察和感知的能力，在经过一定的训练后能计算一些人类不能计算的事情，但是它并没有自适应能力，也就是它不会处理突发的情况，只能处理程序中已经写好的事情，这就叫作弱人工智能。

在 AI 领域，图像识别技术占据着极为重要的地位，而随着计算机技术与信息技术的不断发展，AI 中的图像识别技术的应用范围不断扩展，例如 IBM 的 Watson 医疗诊断、各种指纹识别、各种手机 App 的面部识别，以及百度地图中全景卫星云图识别等都属于这一应用的典型。

AI 技术已经应用于日常生活之中,图像识别技术将来定会有着较为广泛的运用。

1.3.2 机器学习

机器学习(machine learning,ML)是一门多领域交叉学科,涉及概率论、统计学、逼近论、凸分析、算法复杂度理论等多门学科。机器学习专门研究计算机怎样模拟或实现人类的学习行为,以获取新的知识或技能,重新组织已有的知识结构,使之不断改善自身的性能。

机器学习是一类算法的总称,这些算法试图从大量历史数据中挖掘出其中隐含的规律,并用于预测或者分类,更具体地说,机器学习可以看作是寻找一个函数,输入是样本数据,输出是期望的结果,只是这个函数过于复杂,以至于不太方便形式化表达。需要注意的是,机器学习的目标是使学到的函数很好地适用于"新样本",而不仅仅是在训练样本上表现很好。学到的函数适用于新样本的能力,称为泛化(generalization)能力。

机器学习有一个显著的特点,也是机器学习基本的做法,就是使用一个算法从大量的数据中解析并得到有用的信息,并从中学习,然后对之后真实世界中会发生的事情进行预测或作出判断。机器学习需要海量的数据来进行训练,并从这些数据中得到有用的信息,然后反馈到真实世界的用户中。

我们可以用一个简单的例子来说明机器学习:假设在天猫或京东购物的时候,天猫和京东会向我们推送商品信息,这些推荐的商品往往是我们自己很感兴趣的东西,这个过程是通过机器学习完成的。其实这些推送商品是京东和天猫根据我们以前的购物订单和经常浏览的商品记录而得出的结论,可以从中得出商城中的哪些商品是我们感兴趣且有大几率购买的,然后将这些商品定向推送给我们。

1.3.3 深度学习

深度学习(deep learning,DL)是机器学习领域中一个新的研究方向,它被引入机器学习使其更接近于最初的目标——人工智能。深度学习是学习样本数据的内在规律和表示层次,这些学习过程中获得的信息对诸如文字、图像和声音等数据的解释有很大的帮助。它的最终目标是让机器能够像人一样具有分析学习能力,能够识别文字、图像和声音等数据。深度学习是一个复杂的机器学习算法,在语音和图像识别方面取得的效果,远远超过先前相关技术。

深度学习在搜索技术、数据挖掘、机器学习、机器翻译、自然语言处理、多媒体学习、语音、推荐和个性化技术,以及其他相关领域都取得了很多成果。深度学习使机器可以模仿视听和思考等人类的活动,解决了很多复杂的模式识别难题,使得人工智能相关技术取得了很大进步。

1.3.4 基于神经网络的图像识别

神经网络图像识别技术是一种比较新型的图像识别技术,是在传统的图像识别方法和基础上融合神经网络算法的一种图像识别方法。这里的神经网络是指人工神经网络,也就是说这种神经网络并不是动物本身所具有的真正的神经网络,而是人类模仿动物神经网络后人工生成的。在神经网络图像识别技术中,遗传算法与 BP 网络相融合的神经网络图像识别模型是非常经典的,在很多领域都有它的应用。

在图像识别系统中利用神经网络系统，一般会先提取图像的特征，再利用图像所具有的特征映射到神经网络进行图像识别分类。以汽车拍照自动识别技术为例，当汽车通过的时候，汽车自身具有的检测设备会有所感应。此时检测设备就会启用图像采集装置来获取汽车正反面的图像。获取了图像后必须将图像上传到计算机进行保存以便识别。最后车牌定位模块就会提取车牌信息，对车牌上的字符进行识别并显示最终的结果。在对车牌上的字符进行识别的过程中就用到了基于模板匹配算法和基于人工神经网络算法。

1.3.5 基于非线性降维的图像识别

计算机的图像识别技术是一个高维的识别技术，不管图像本身的分辨率如何，其产生的数据经常是多维性的，这给计算机的识别带来了非常大的困难。想让计算机具有高效的识别能力，最直接有效的方法就是降维。降维分为线性降维和非线性降维。例如，主成分分析（PCA）和线性奇异分析（LDA）等就是常见的线性降维方法，它们的特点是简单、易于理解。但是通过线性降维处理的是整体的数据集合，所求的是整个数据集合的最优低维投影。

经过验证，这种线性的降维策略计算复杂度高而且占用相对较多的时间和空间，因此就产生了基于非线性降维的图像识别技术，它是一种极其有效的非线性特征提取方法。此技术可以发现图像的非线性结构而且可以在不破坏其本征结构的基础上对其进行降维，使计算机的图像识别在尽量低的维度上进行，这样就提高了识别速率。例如，人脸图像识别系统所需的维数通常很高，其复杂度之高对计算机来说无疑是巨大的"灾难"。由于在高维度空间中人脸图像的不均匀分布，使得人类可以通过非线性降维技术来得到分布紧凑的人脸图像，从而提高人脸识别技术的高效性。

总之，随着深度学习和计算机硬件的发展，特别是卷积神经网络的出现，图像识别技术取得了巨大的进步。现在的图像识别系统在许多任务上已经超越了人类的表现，并且在许多领域获得了广泛的应用。

第 2 章　图像的采样、变换和卷积处理

在图像处理和计算机视觉领域，采样、变换和卷积处理是常用的图像处理操作，它们在不同的任务中起着重要的作用。本章详细讲解使用 Python 语言实现采样、变换和卷积处理的知识。

2.1　采样

采样是指从原始图像中选择一部分像素来表示图像的过程，常见的采样方法包括降采样和上采样。

- 降采样（downsampling）：降采样是减少图像分辨率的过程，通过从原始图像中选择部分像素来构建低分辨率图像。降采样可以用于图像压缩、缩小等应用。
- 上采样（upsampling）：上采样是增加图像分辨率的过程，通过插值等方法在原始图像的像素之间插入新的像素来构建高分辨率图像。上采样可以用于图像放大、重建等应用。

2.1.1　最近邻插值采样

最近邻插值（nearest neighbor interpolation）是一种简单的采样方法，它将目标像素的值设置为最接近它的原始像素的值。在 Python 程序中，有多种实现最近邻插值采样的方法，下面是几种常见的方法。

1．使用 PIL 库

PIL 库是一种常用的图像处理库，通过其内置函数 resize() 可以指定插值方法为最近邻插值进行图像采样，可以使用 Image.NEAREST 参数来指定最近邻插值方法。例如，在下面的实例文件 jin.py 中，演示了使用 PIL 库实现最近邻插值采样的过程。

实例 2-1：使用 PIL 库实现最近邻插值采样

源码路径： daima\2\jin.py

```python
from PIL import Image

# 打开图像
image = Image.open('111.jpg')

# 定义目标尺寸
target_size = (800, 600)

# 进行最近邻插值采样
resized_image = image.resize(target_size, Image.NEAREST)

# 显示采样后的图像
resized_image.show()
```

2. 使用 scikit-image 库

scikit-image 库也提供了最近邻插值的函数，可以方便地进行图像采样操作。我们可以使用函数 skimage.transform.resize()进行采样，将参数 order 设置为 0 来指定最近邻插值。例如，在下面的实例文件 sjin.py 中，演示了使用库 scikit-image 实现最近邻插值采样的过程。

实例 2-2：使用 scikit-image 库实现最近邻插值采样

源码路径：daima\2\sjin.py

```
from skimage import io, transform

# 读取图像
image = io.imread('111.jpg')

# 定义目标尺寸
target_size = (800, 600)

# 进行最近邻插值采样
resized_image = transform.resize(image, target_size, order=0)

# 显示采样后的图像
io.imshow(resized_image)
io.show()
```

3. 使用 NumPy 库

我们也可以使用 NumPy 库实现最近邻插值采样功能，通过索引操作和取整操作来实现。这种方法需要手动计算目标像素在原始图像中的位置，并取最近的像素值。例如，在下面的实例文件 njin.py 中，演示了使用 NumPy 库实现最近邻插值采样的过程。

实例 2-3：使用 NumPy 库实现最近邻插值采样

源码路径：daima\2\njin.py

```
import numpy as np
from skimage import io

# 读取图像
image = io.imread('input.jpg')

# 定义目标尺寸
target_size = (800, 600)

# 计算采样比例
scale_x = image.shape[0] / target_size[0]
scale_y = image.shape[1] / target_size[1]

# 构建目标图像
resized_image = np.zeros((target_size[0], target_size[1], image.shape[2]), dtype=np.uint8)

# 遍历目标图像的每个像素
for i in range(target_size[0]):
    for j in range(target_size[1]):
        # 计算原始图像中的位置
        x = int(i * scale_x)
        y = int(j * scale_y)
```

```
            # 最近邻插值
            resized_image[i, j, :] = image[x, y, :]

# 显示采样后的图像
io.imshow(resized_image)
io.show()
```

对上述代码的具体说明如下：

（1）使用函数 skimage.io.imread()读取输入图像，并将其存储在 image 变量中。然后，定义了目标尺寸（target_size）。

（2）计算采样比例（scale_x 和 scale_y），通过将原始图像的维度除以目标尺寸的对应维度来计算得到。

（3）使用函数 np.zeros()创建了一个与目标尺寸和原始图像通道数相匹配的空白图像。

（4）使用双层循环遍历目标图像的每个像素，并根据最近邻插值的原理从原始图像中选择最接近的像素值赋给目标图像的对应位置。

（5）使用函数 skimage.io.imshow()显示采样后的图像。

2.1.2 双线性插值

双线性插值（bilinear interpolation）是一种基于四个最近邻像素的插值方法，它使用线性加权平均来计算目标像素的值。在 Python 程序中，实现双线性插值采样的方法有如下几种：

1. 使用 OpenCV 库

在 OpenCV 库中提供了内置函数 cv2.resize()，通过此函数设置插值方法为 cv2.INTER_LINEAR，这样可以实现双线性插值采样功能。例如，在下面的实例文件 cvshuang.py 中，演示了使用 OpenCV 库实现双线性插值采样的过程。

实例 2-4：使用 OpenCV 库实现双线性插值采样

源码路径： daima\2\cvshuang.py

```
import cv2
from skimage import io

# 读取图像
image = io.imread('input.jpg')

# 定义目标尺寸
target_size = (800, 600)

# 使用 OpenCV 进行双线性插值采样
resized_image = cv2.resize(image, target_size, interpolation=cv2.INTER_LINEAR)

# 显示采样后的图像
io.imshow(resized_image)
io.show()
```

在上述代码中，使用函数 cv2.resize()对图像进行双线性插值采样。通过指定目标尺寸和插值方法 cv2.INTER_LINEAR，可以得到双线性插值后的图像。

2. 使用 SciPy 库

SciPy 库中的函数 scipy.ndimage.zoom()可以进行双线性插值采样，通过指定参数 order=1 可以使用双线性插值方法。下面是一个使用函数 scipy.ndimage.zoom()进行双线性插值采样的例子。

实例 2-5：将指定图像放大两倍
源码路径：daima\2\scshuang.py

```python
import numpy as np
import matplotlib.pyplot as plt
from scipy import ndimage

# 生成一个简单的图像
image = np.zeros((5, 5))
image[2, 2] = 1

# 双线性插值采样，放大两倍
zoomed_image = ndimage.zoom(image, 2, order=1)

# 绘制原始图像和采样后的图像
plt.subplot(1, 2, 1)
plt.title('Original Image')
plt.imshow(image, cmap='gray')

plt.subplot(1, 2, 2)
plt.title('Zoomed Image')
plt.imshow(zoomed_image, cmap='gray')

plt.show()
```

在上述代码中，首先生成一个简单的图像，其中心像素的值为 1，其余像素的值为 0。然后，我们使用 ndimage.zoom()函数将图像放大两倍，这一步通过指定放大倍数为 2 来实现。使用 order=1 参数来指定双线性插值方法。最后，我们使用 Matplotlib 库绘制原始图像和采样后的图像。运行上述代码，将看到原始图像和双线性插值采样后的图像。双线性插值采样通过对原始像素周围的四个像素进行加权平均来计算新像素的值，从而实现了放大效果。代码执行后的效果如图 2-1 所示。

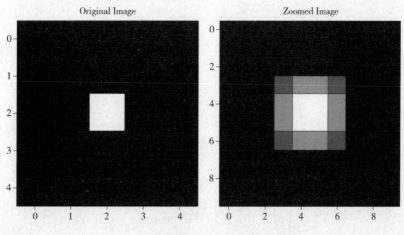

图 2-1　使用 SciPy 库将图像放大两倍

注意：在实际应用中，可以将 image 替换为自己的图像数据，通过调整放大倍数和其他参数来满足自己的需求。

3. 使用 NumPy 库

虽然在 NumPy 库中没有直接提供双线性插值的函数，但是可以通过自定义编程方法实现。例如，下面是一个使用 NumPy 库实现双线性插值采样的实用例子。

实例 2-6：使用 NumPy 库实现双线性插值采样（将指定图像放大两倍）

源码路径：daima\2\nshuang.py

```python
import numpy as np
import matplotlib.pyplot as plt

def bilinear_interpolation(image, zoom_factor):
    # 原始图像尺寸
    height, width = image.shape
    
    # 目标图像尺寸
    new_height = int(height * zoom_factor)
    new_width = int(width * zoom_factor)
    
    # 生成目标图像
    zoomed_image = np.zeros((new_height, new_width))
    
    # 计算插值权重
    y_ratio = height / new_height
    x_ratio = width / new_width
    
    for i in range(new_height):
        for j in range(new_width):
            # 计算在原始图像中的坐标
            y = i * y_ratio
            x = j * x_ratio
            
            # 计算相邻四个像素的索引
            x1 = int(np.floor(x))
            x2 = min(x1 + 1, width - 1)
            y1 = int(np.floor(y))
            y2 = min(y1 + 1, height - 1)
            
            # 计算插值权重
            dx = x - x1
            dy = y - y1
            
            # 双线性插值
            interpolated_value = (1 - dx) * (1 - dy) * image[y1, x1] + \
                                 dx * (1 - dy) * image[y1, x2] + \
                                 (1 - dx) * dy * image[y2, x1] + \
                                 dx * dy * image[y2, x2]
            
            # 将插值结果放入目标图像
            zoomed_image[i, j] = interpolated_value
    
    return zoomed_image
```

```python
# 生成一个简单的图像
image = np.zeros((5, 5))
image[2, 2] = 1

# 双线性插值采样,放大两倍
zoomed_image = bilinear_interpolation(image, 2)

# 绘制原始图像和采样后的图像
plt.subplot(1, 2, 1)
plt.title('Original Image')
plt.imshow(image, cmap='gray')

plt.subplot(1, 2, 2)
plt.title('Zoomed Image')
plt.imshow(zoomed_image, cmap='gray')

plt.show()
```

在上述代码中,定义了一个 bilinear_interpolation()函数来实现双线性插值采样。函数中首先计算目标图像的尺寸,并生成一个空白的目标图像。然后,通过嵌套的循环遍历目标图像的每个像素,并计算在原始图像中的坐标。接下来,通过计算相邻四个像素的索引和插值权重,使用双线性插值公式计算新像素的值,并将其放入目标图像中。最后,使用 Matplotlib 库绘制原始图像和采样后的图像。

运行上述代码,将看到原始图像和双线性插值采样后的图像。双线性插值采样通过对原始像素周围的四个像素进行加权平均来计算新像素的值,从而实现了放大效果。代码执行后的效果如图 2-2 所示。

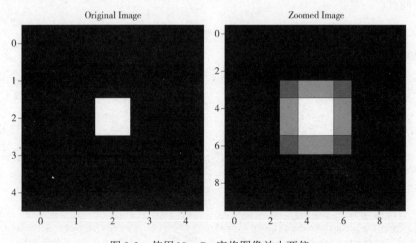

图 2-2　使用 NumPy 库将图像放大两倍

2.1.3　双立方插值

双立方插值(bicubic interpolation)是一种更为精确的插值方法,它使用 16 个最近邻像素进行计算。在 Python 程序中,可以使用下面的方法实现双立方插值采样操作:

1. 使用 OpenCV 库

在 OpenCV 库中提供了函数 cv2.resize(),通过指定插值方法为 cv2.INTER_CUBIC 可以实现双立方插值采样。例如，在下面的实例文件 cvli.py 中，演示了使用库 OpenCV 实现双立方插值采样的过程。

实例 2-7：使用 OpenCV 库实现双立方插值采样

源码路径：daima\2\cvli.py

```python
import cv2
from skimage import io

# 读取图像
image = io.imread('input.jpg')

# 定义目标尺寸
target_size = (800, 600)

# 使用OpenCV进行双立方插值采样
resized_image = cv2.resize(image, target_size, interpolation=cv2.INTER_CUBIC)

# 显示采样后的图像
io.imshow(resized_image)
io.show()
```

在上述代码中，使用 cv2.resizeo()函数对图像进行双立方插值采样。通过指定目标尺寸和插值方法 cv2.INTER_CUBIC，可以得到双立方插值后的图像。

2．使用 SciPy 库

SciPy 库中的函数 scipy.ndimage.zoom()可以进行双立方插值采样，通过指定参数"order=3"可以使用双立方插值方法。例如，在下面的实例文件 scli.py 中，演示了使用库 SciPy 实现双立方插值采样的过程。

实例 2-8：使用 SciPy 库实现双立方插值采样

源码路径：daima\2\scli.py

```python
import numpy as np
from scipy import ndimage
from skimage import io

# 读取图像
image = io.imread('111.jpg')

# 定义目标尺寸
target_size = (800, 600)

# 使用SciPy进行双立方插值采样
resized_image = ndimage.zoom(image, (target_size[0]/image.shape[0], target_size[1]/image.shape[1], 1), order=3)

# 显示采样后的图像
io.imshow(resized_image)
io.show()
```

在上述代码中，使用函数 ndimage.zoom()对图像进行双立方插值采样。通过计算目标尺寸与原始图像尺寸的比例，并传入函数 zoom()进行采样。使用参数 order=3 指定双立方插值方法。

3．使用 PIL 库

在 PIL 库中提供了函数 Image.resize()，通过指定参数 resample=Image.BICUBIC 可以实现双立

方插值采样。例如，在下面的实例文件 pili.py 中，演示了使用 PIL 库实现双立方插值采样的过程。

实例 2-9：使用 PIL 库实现双立方插值采样

源码路径：daima\2\plli.py

```python
from PIL import Image
from skimage import io

# 读取图像
image = Image.open('input.jpg')

# 定义目标尺寸
target_size = (800, 600)

# 使用 PIL 进行双立方插值采样
resized_image = image.resize(target_size, resample=Image.BICUBIC)

# 显示采样后的图像
io.imshow(resized_image)
io.show()
```

在上述代码中，使用函数 Image.resize()对图像进行双立方插值采样。通过指定目标尺寸和参数 resample=Image.BICUBIC，可以实现双立方插值采样。

注意：由于双立方插值是一种计算量较大的插值方法，所以在实际应用中需要权衡计算速度和插值效果。

2.1.4 lanczos 插值

lanczos 插值（lanczos interpolation）是一种使用窗口函数的插值方法，它在保持图像细节的同时进行平滑。在 Python 程序中，可以通过如下方法实现 lanczos 插值采样功能：

1．使用 PIL 库

在 PIL 库中提供了函数 Image.resize()，通过指定参数 resample=Image.LANCZOS 可以实现 lanczos 插值采样功能。例如，在下面的实例文件 picha.py 中，演示了使用 PIL 库实现 lanczos 插值采样的过程。

实例 2-10：使用 PIL 库实现 lanczos 插值采样

源码路径：daima\2\picha.py

```python
from PIL import Image
from skimage import io

# 读取图像
image = Image.open('111.jpg')

# 定义目标尺寸
target_size = (800, 600)

# 使用 PIL 进行 Lanczos 插值采样
resized_image = image.resize(target_size, resample=Image.LANCZOS)

# 显示采样后的图像
io.imshow(resized_image)
io.show()
```

在上述代码中，使用函数 Image.resize()设置指定目标尺寸和参数 resample=Image.LANCZOS，实现 lanczos 插值采样功能。值得注意的是，lanczos 插值是一种计算量较大的插值方法，适用于放大图像的情况。在实际应用中，需要权衡计算速度和插值效果，选择合适的插值方法来满足需求。

2．使用 SciPy 库

在 SciPy 库中提供了函数 scipy.ndimage.zoom()来进行 lanczos 插值采样。通过指定 order=0 和参数 mode=nearest 可以使用 lanczos 插值方法。例如，在下面的实例文件 sccha.py 中，演示了使用库 SciPy 实现 lancz os 插值采样的过程。

实例 2-11：使用 SciPy 库实现 lanczos 插值采样

源码路径：daima\2\sccha.py

```python
import numpy as np
from scipy import ndimage
from skimage import data, io

# 读取图像
image = data.camera()

# 定义放大倍数
scale_factor = 2

# 计算目标尺寸
target_shape = (image.shape[0] * scale_factor, image.shape[1] * scale_factor)

# 使用 SciPy 进行插值采样
resized_image = ndimage.zoom(image, scale_factor, order=3)

# 显示采样后的图像
io.imshow(resized_image, cmap='gray')
io.show()
```

对上述代码的具体说明如下：
- 使用 data.camera()函数生成了一个示例图像。然后，定义了放大倍数 scale_factor，这里设置为 2。通过计算目标尺寸 target_shape，即原始图像尺寸乘以放大倍数，确定了采样后的图像大小。
- 使用 ndimage.zoom()函数对图像进行插值采样。在本例中，使用了 order=3 参数表示使用双立方插值方法。
- 使用函数 skimage.io.imshow()显示采样后的图像，并通过设置 cmap=gray 来指定灰度色彩映射。

2.2　离散傅里叶变换（DFT）

离散傅里叶变换（discrete fourier transform，DFT）是一种将离散信号转换为频域表示的数学技术，它通过将信号分解为一系列正弦和余弦函数的和来表示信号的频谱特征。

2.2.1　为什么使用 DFT

离散傅里叶变换将输入信号从时域转换到频域，得到信号在不同频率上的幅度和相位信息。

通过进行频域分析，我们可以提取信号的频谱特征，例如频率成分、频率强度、相位关系等。

DFT 在信号处理和频域分析中具有广泛的应用，下面是几个使用 DFT 的主要原因：

- 频域表示。DFT 将信号从时域转换到频域，通过分析信号在不同频率上的成分，可以获取信号的频谱信息。频域表示可以帮助我们理解信号的频率特性、频率成分之间的相互关系，以及信号中存在的噪声或干扰。
- 频谱分析。DFT 可以用于频谱分析，通过分析信号的频谱可以提取信号的频率成分、频率强度及频域特征。这对于识别信号中的特定频率成分、峰值频率、频率分布等非常有用，例如音频处理中的音调识别、信号处理中的滤波和频谱估计等。
- 信号滤波。在频域中，可以对信号进行滤波操作。通过将频域中的特定频率成分滤除或增强，可以实现信号的频域滤波。DFT 可以用于设计和实现各种滤波器，例如低通滤波器、高通滤波器、带通滤波器等。
- 压缩和编码。DFT 可以应用于图像和音频的压缩和编码。通过在频域中对信号进行表示，可以利用信号的频率特性进行数据压缩和信息编码，以实现更高效的数据存储和传输。
- 信号重构。离散傅里叶逆变换（IDFT）可以将频域信号转换回时域。这对于从频域信号中恢复原始时域信号或合成新的信号非常有用，例如音频合成和图像生成等应用。

总之，DFT 是一种强大的数学工具，可以帮助我们理解信号的频率特性、频域成分和频谱信息。它在信号处理、频谱分析、滤波、压缩编码及信号重构等方面发挥着重要的作用，并在许多领域中得到广泛应用。

2.2.2 使用 NumPy 库实现 DFT

在 NumPy 库中，函数 numpy.fft.fft()用于计算一维和多维离散傅里叶变换，该函数接收一个输入信号，并返回对应的频域表示。例如，下面是一个使用 NumPy 库实现 DFT 的例子，功能是将一个正弦波信号进行离散傅里叶变换。

实例 2-12：将一个正弦波信号进行离散傅里叶变换

源码路径： daima\2\ndft.py

```python
import numpy as np
import matplotlib.pyplot as plt

# 定义正弦波信号
t = np.linspace(0, 1, 1000)
x = np.sin(2 * np.pi * 10 * t)

# 进行离散傅里叶变换
X = np.fft.fft(x)

# 计算频率轴
freq = np.fft.fftfreq(len(x))

# 绘制频域表示
plt.plot(freq, np.abs(X))
plt.xlabel('Frequency')
plt.ylabel('Amplitude')
```

```
plt.title('Discrete Fourier Transform')
plt.show()
```

在上述代码中生成了一个频率为 10 Hz 的正弦波信号，并使用函数 np.fft.fft()进行离散傅里叶变换。然后，通过计算频率轴 freq，可以绘制出信号的频域表示。代码执行后的效果如图 2-3 所示。

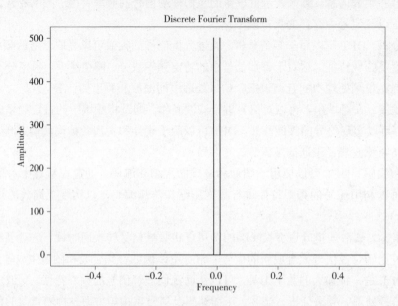

图 2-3　使用 NumPy 库对一个正弦波信号实现 DFT

2.2.3　使用 SciPy 库实现 DFT

在 SciPy 库中提供了一个独立的 fft 模块，用于计算离散傅里叶变换。模块 fft 提供了多个关于傅里叶变换的功能和选项。例如，下面是一个使用 SciPy 库的例子，功能是将一个信号进行离散傅里叶变换并绘制频谱图。

实例 2-13：将一个信号进行离散傅里叶变换并绘制频谱图

源码路径：daima\2\scdft.py

```
import numpy as np
import matplotlib.pyplot as plt
from scipy.fft import fft

# 定义输入信号
x = np.random.rand(100)

# 进行离散傅里叶变换
X = fft(x)

# 计算频率轴
freq = np.fft.fftfreq(len(x))

# 绘制频谱图
plt.plot(freq, np.abs(X))
plt.xlabel('Frequency')
plt.ylabel('Amplitude')
plt.title('Discrete Fourier Transform using SciPy')
plt.show()
```

在上述代码中生成了一个随机信号,并使用 fft()函数进行离散傅里叶变换。然后,通过计算频率轴 freq 绘制出信号的频谱图。代码执行后的效果如图 2-4 所示。

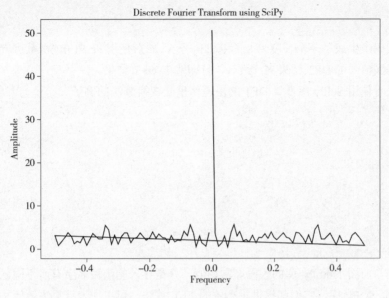

图 2-4　使用 SciPy 库实现 DFT

2.2.4　用快速傅里叶变换算法计算 DFT

快速傅里叶变换(fast fourier transform,FFT)是一种高效计算离散傅里叶变换的算法。FFT 和 DFT 都是用于计算信号的频域表示的方法,但它们在计算效率和算法实现上有所不同。

1. 计算效率
- DFT 是通过直接计算定义式来获得频域表示,其计算复杂度为 $O(N^2)$,其中 N 是信号的长度。
- FFT 是一种基于分治思想的快速算法,通过将 DFT 的计算复杂度从 $O(N^2)$ 降低到 $O(N\log_2 N)$。FFT 算法利用了信号的对称性和周期性,将 DFT 的计算任务分解为一系列较小的 DFT 计算。

2. 算法实现
- DFT 是一种直接的计算方法,按照定义式进行求和运算,需要遍历信号的所有时间点和频率点,计算量较大。
- FFT 是一种基于蝶形运算的迭代算法,通过递归和迭代的方式对信号进行分解和合成,减少了计算量。

总结来说,FFT 是一种通过巧妙的算法设计和优化实现的快速计算 DFT 的方法。相对于 DFT,FFT 算法在计算复杂度上具有更高的效率,特别适用于处理大规模数据和实时信号处理任务。

在 Python 程序中,可以使用 NumPy 库中的函数 numpy.fft.fft()或 SciPy 库中的函数 scipy.fft.fft()实现 FFT 计算。例如,下面是使用 NumPy 库计算 DFT 的快速傅里叶变换算法的例子。

```
import numpy as np

# 定义输入信号
x = np.random.rand(8)
```

```
# 使用FFT计算DFT
X = np.fft.fft(x)

# 打印DFT结果
print("DFT 结果: ", X)
```

在上述代码中生成了一个长度为 8 的随机信号 X，然后使用 np.fft.fft()函数进行快速傅里叶变换计算，得到信号的 DFT 结果 X。最后，打印输出 DFT 结果。

下面再来看使用 SciPy 库计算 DFT 的快速傅里叶变换算法的例子。

```
import numpy as np
from scipy.fft import fft

# 定义输入信号
x = np.random.rand(8)

# 使用FFT计算DFT
X = fft(x)

# 打印DFT结果
print("DFT 结果: ", X)
```

上述代码与前面 NumPy 库的代码非常相似，只是导入的函数名字有所不同。无论是使用 NumPy 库还是 SciPy 库，它们都提供了高效的 FFT 算法，可以快速计算离散傅里叶变换，对于频域分析和信号处理任务非常有用。

注意：在实际应用中，通常使用 FFT 而不是直接使用 DFT 来计算信号的频域表示，因为 FFT 能够以更短的计算时间提供相同的结果。不过需要注意的是，FFT 在计算过程中会对信号进行周期延拓，因此在应用中需要注意信号的边界处理和频谱的解释。

2.3 卷积

卷积是一种数学运算，用于信号处理、图像处理和其他领域中。在信号和图像处理中，卷积常用于信号的滤波、特征提取和图像处理等任务。

2.3.1 为什么需要卷积图像

卷积在图像处理中起着重要的作用，可以用于多种图像处理任务。卷积在图像处理中的常见功能如下：

- 模糊和平滑：通过应用卷积核，可以对图像进行模糊和平滑处理，以降低图像中的噪声，并使图像变得更加平滑。常见的卷积核包括均值滤波器和高斯滤波器。
- 锐化和边缘检测：卷积核可以用于提取图像中的边缘和细节信息。通过应用一些特定的卷积核，如 sobel、prewitt 和 laplacian 卷积核，可以增强图像中的边缘，并使其更加清晰和突出。
- 图像增强：卷积可以用于增强图像的特定特征。例如，可以使用卷积核来增强图像的纹理、对比度或颜色饱和度，以改善图像的视觉效果。
- 特征提取：卷积在图像识别和计算机视觉中被广泛用于特征提取。通过将图像与一组预定义的卷积核进行卷积运算，可以提取出图像中的各种特征，如边缘、角点、纹理等。这些特征可以用于对象检测、图像分类和图像分割等任务。
- 图像重建和去噪：通过卷积运算，可以将模糊、噪声或低分辨率图像恢复为更清晰、更高质量的图像。例如，可以使用反卷积技术对模糊图像进行逆滤波，或使用卷积核进行图像超分辨率重建。

上述功能仅是卷积在图像处理中的一些常见应用，卷积还有很多其他的用途，具体取决于所使用的卷积核和处理目标。通过调整卷积核的参数和选择不同的卷积核，可以实现不同的图像处理效果。

2.3.2 使用 SciPy 库中的函数 convolve2d() 进行卷积操作

在 Python 程序中，可以使用 SciPy 库中的函数 convolve2d() 对二维图像实现卷积操作。函数 convolve2d() 的语法格式如下：

```
scipy.signal.convolve2d(in1, in2, mode='full', boundary='fill', fillvalue=0)
```

- in1：输入数组，表示要进行卷积操作的第一个输入。
- in2：卷积核，表示要进行卷积操作的第二个输入。
- mode：卷积模式，默认为 full，表示输出的大小与输入数组的大小相同。其他可选值为 valid 和 same，分别表示输出大小将根据输入数组和卷积核的大小进行调整。
- boundary：边界处理方式，默认为 fill，表示使用填充值进行边界处理。其他可选值为 wrap 和 symm，分别表示使用循环填充和对称填充进行边界处理。
- fillvalue：填充值，在 boundary=fill 时使用，用于指定边界填充的数值。
- 返回值：返回卷积操作的结果数组。

在使用函数 convolve2d() 时，in1 和 in2 可以是二维数组、二维图像或多通道图像。在进行卷积操作时，输入数组和卷积核进行逐元素相乘，并将乘积结果求和得到卷积结果。卷积操作可以用于信号处理、图像处理、卷积神经网络等领域。例如，下面是一个使用函数 convolve2d() 进行卷积操作的例子，对 skimage 库中的内置素材图像 camera 进行卷积操作。

实例 2-14：对 skimage 库中的内置素材图像 camera 进行卷积操作

源码路径：daima\2\con1.py

```
import numpy as np
from scipy import signal
import matplotlib.pyplot as plt
from skimage import data
# 设置中文字体
plt.rcParams["font.sans-serif"] = "SimHei"
plt.rcParams["axes.unicode_minus"] = False
# 读取图像
image = data.camera()
kernel = np.array([[0, -1, 0],
                   [-1, 5, -1],
                   [0, -1, 0]])

# 执行卷积操作
result = signal.convolve2d(image, kernel, mode='same')
```

```
# 可视化输入图像和卷积结果
fig, axes = plt.subplots(1, 2, figsize=(10, 5))
axes[0].imshow(image, cmap='gray')
axes[0].set_title('输入图像')

axes[1].imshow(result, cmap='gray')
axes[1].set_title('卷积结果')

plt.tight_layout()
plt.show()
```

在本实例中，卷积核采用了一个简单的边缘检测算子。执行卷积操作后，通过 Matplotlib 库将输入图像和卷积结果进行可视化展示。代码执行后的效果如图 2-5 所示。

图 2-5 使用函数 convolve2d()对图像进行卷积操作

2.3.3 使用 SciPy 库中的函数 ndimage.convolve()进行卷积操作

在 Python 程序中，还可以使用 SciPy 库中的函数 ndimage.convolve()进行卷积操作。函数 ndimage.convolve()的语法格式如下：

```
scipy.ndimage.convolve(input, weights, output=None, mode='reflect', cval=0.0, origin=0)
```

- input：输入数组，表示要进行卷积操作的数组。
- weights：卷积核，表示要进行卷积操作的权重数组。
- output：输出数组，可选参数，用于存储卷积结果。如果未提供，则会创建一个新的数组来存储结果。
- mode：边界处理模式，默认为 reflect，表示使用镜像反射模式进行边界处理。其他可选值包括 constant、nearest、mirror 和 wrap，分别表示常数扩展、最近邻扩展、镜像反射扩展和循环扩展。
- cval：当 mode=constant 时，用于指定常数扩展模式下的填充值。
- origin：卷积核的原点位置，默认为(0, 0)，表示卷积核的中心位置。
- 返回值：返回卷积操作的结果数组。

函数 ndimage.convolve()可以对任意维度的数组进行卷积操作，常应用于图像处理、信号处理等领域。例如，下面是一个使用函数 scipy.ndimage.convolve()进行图像卷积的例子，使用了一张具体的图像 111.jpg 作为输入,并采用边界模式为 mirror。

实例 2-15:对指定的图像进行卷积操作

源码路径: daima\2\ndi.py

```python
import numpy as np
import matplotlib.pyplot as plt
from scipy import ndimage

# 读取图像
image = plt.imread('111.jpg')

# 将图像转换为灰度图
image_gray = np.mean(image, axis=2)

# 定义卷积核
kernel = np.array([[0, -1, 0],
                   [-1, 5, -1],
                   [0, -1, 0]])

# 执行卷积操作
result = ndimage.convolve(image_gray, kernel, mode='mirror')

# 显示原始图像和卷积结果
plt.subplot(1, 2, 1)
plt.imshow(image_gray, cmap='gray')
plt.title('Original Image')

plt.subplot(1, 2, 2)
plt.imshow(result, cmap='gray')
plt.title('Convolved Image')

plt.show()
```

在上述代码中,首先使用函数 np.mean() 将彩色图像转换为灰度图像,以便与定义的卷积核进行匹配。然后,使用函数 ndimage.convolve() 对灰度图像进行卷积操作,并将结果存储在 result 变量中。最后,我们使用函数 plt.imshow() 显示原始图像和卷积结果。代码执行后的效果如图 2-6 所示。

图 2-6 使用函数 ndimage.convolve() 对图像进行卷积操作

2.4 频域滤波

频域滤波是一种基于信号的频谱进行处理的滤波方法。它通过将信号转换到频域，应用滤波操作，然后再将信号转换回时域，以实现滤波效果。

2.4.1 什么是滤波器

滤波器是信号处理中常用的工具，用于对信号进行频率选择或干扰去除。它通过对输入信号进行加权求和或乘积运算，改变信号的频谱特性或时间域特性，从而实现对信号的调整、增强或去除干扰等目的。

在图像处理中，滤波器通常用于平滑图像、增强图像细节、边缘检测、去噪等应用。滤波器可以基于不同的原理和方法进行设计，常见的滤波器类型包括低通滤波器、高通滤波器、带通滤波器、带阻滤波器等。

滤波器可以在时域或频域中操作。在时域中，滤波器通过对输入信号的每个采样点应用滤波算法来改变信号的特性。常见的时域滤波器包括移动平均滤波器、中值滤波器和高斯滤波器等。在频域中，滤波器通过对输入信号的傅里叶变换进行操作来改变信号的频谱特性。常见的频域滤波器包括傅里叶变换滤波器和小波变换滤波器等。

滤波器的设计和选择取决于应用的需求和目标。不同类型的滤波器可以对信号进行不同的处理和调整，从而实现不同的信号处理目标。滤波器设计涉及到信号处理理论和技术，需要考虑信号的特性、噪声情况、滤波器的频率响应等因素。

在 Python 程序中，常用的库如 SciPy 和 OpenCV 提供了丰富的滤波器函数和工具，可以方便地进行滤波操作。这些库提供了各种类型的滤波器，可以根据具体需求选择适合的滤波器进行信号处理。

2.4.2 高通滤波器

高通滤波器是一种用于增强图像中高频信息或减弱低频信息的滤波器。它可以帮助我们突出图像中的细节、边缘和纹理等高频特征，同时抑制图像中的低频部分，例如平坦区域和背景等。

在频域中，高通滤波器可以通过在频谱中去除低频分量来实现。常见的高通滤波器包括布特沃斯高通滤波器、高斯高通滤波器和锐化滤波器等。例如，下面是一个使用 SciPy 库进行高通滤波的例子，使用函数 ndimage.gaussian_filter()时对图像应用了高斯高通滤波器，并指定了一个特定的 sigma 值。然后，从原始图像中减去平滑后的图像，得到滤波后的图像。

实例 2-16：对图像应用高斯高通滤波器并得到滤波后的图像

源码路径： daima\2\gaolv.py

```python
import numpy as np
from scipy import ndimage
import matplotlib.pyplot as plt

# 加载图像
image = plt.imread('111.jpg')

# 如果需要，将图像转换为灰度图像
if len(image.shape) > 2:
```

```
    image = np.mean(image, axis=2)

# 对图像应用高斯滤波器以平滑图像
smoothed_image = ndimage.gaussian_filter(image, sigma=5)

# 从原始图像中减去平滑后的图像实现高通滤波器
high_pass_image = image - smoothed_image

# 显示原始图像和高通滤波后的图像
fig, axes = plt.subplots(1, 2, figsize=(10, 5))
axes[0].imshow(image, cmap='gray')
axes[0].set_title('原始图像')
axes[0].axis('off')
axes[1].imshow(high_pass_image, cmap='gray')
axes[1].set_title('高通滤波后的图像')
axes[1].axis('off')
plt.show()
```

在上述代码中，首先读取一张彩色图像，并将其转换为灰度图像。对灰度图像进行傅里叶变换，得到频谱表示。接下来，创建一个高斯高通滤波器，其中参数 cutoff_freq 控制滤波的频率截断点。然后，将滤波器应用于频谱，得到滤波后的频谱。最后，使用逆傅里叶变换将滤波后的频谱转换回时域，得到滤波后的图像。代码执行后的效果如图 2-7 所示。

图 2-7　对图像应用高斯高通滤波器后的效果

2.4.3　低通滤波器

低通滤波器是一种用于图像处理的滤波器，它允许通过较低频率的信号而抑制高频信号。低通滤波器在图像处理中常用于平滑图像、去噪和模糊化等。在 Python 程序中，可以使用 SciPy 库中的模块 ndimage 来实现低通滤波器。其中，函数 gaussian_filter()可以用于应用高斯模糊操作，从而实现低通滤波器的效果。例如，下面是一个使用模块 ndimage 函数 gaussian_filter()实现低通滤波器的例子。

实例 2-17：使用函数 gaussian_filter()实现低通滤波器
源码路径：daima\2\dilv.py

```
import numpy as np
from scipy import ndimage
import matplotlib.pyplot as plt
# 设置中文字体
plt.rcParams["font.sans-serif"] = "SimHei"
plt.rcParams["axes.unicode_minus"] = False

# 加载图像
image = plt.imread('888.jpg')

# 如果需要，将图像转换为灰度图像
if len(image.shape) > 2:
```

```python
        image = np.mean(image, axis=2)

# 应用高斯模糊操作实现低通滤波器
blurred_image = ndimage.gaussian_filter(image, sigma=3)

# 显示原始图像和低通滤波后的图像
fig, axes = plt.subplots(1, 2, figsize=(10, 5))
axes[0].imshow(image, cmap='gray')
axes[0].set_title('原始图像')
axes[0].axis('off')
axes[1].imshow(blurred_image, cmap='gray')
axes[1].set_title('低通滤波后的图像')
axes[1].axis('off')
plt.show()
```

在上述代码中，使用函数 ndimage.gaussian_filter()将一个具有特定 sigma 值的高斯模糊应用于图像。通过调整 sigma 值，可以控制模糊程度，从而实现不同程度的低通滤波效果。代码执行后的效果如图 2-8 所示。

图 2-8　使用函数 gaussian_filter()实现的效果

2.4.4　DoG 带通滤波器

DoG（difference of gaussians）是一种带通滤波器，它是通过对不同尺度的高斯滤波器之间的差异进行计算而得到的。DoG 滤波器可以用于图像处理中的边缘检测、特征提取和纹理分析等任务。

在 Python 程序中，可以使用 SciPy 库中的模块 ndimage 实现 DoG 带通滤波器。其中，函数 gaussian_filter()用于应用高斯模糊操作，函数 difference()用于计算两个高斯模糊图像之间的差异。例如，下面是一个使用模块 ndimage 实现 DoG 带通滤波器的例子。

实例 2-18：使用模块 ndimage 实现 DoG 带通滤波器

源码路径： daima\2\dailv.py

```
import numpy as np
from scipy import ndimage
import matplotlib.pyplot as plt
```

```python
# 设置中文字体
plt.rcParams["font.sans-serif"] = "SimHei"
plt.rcParams["axes.unicode_minus"] = False
# 加载图像
image = plt.imread('888.jpg')

# 如果需要，将图像转换为灰度图像
if len(image.shape) > 2:
    image = np.mean(image, axis=2)

# 定义两个不同尺度的高斯模糊参数
sigma1 = 2.0
sigma2 = 5.0

# 应用高斯模糊操作获取两个模糊图像
blurred1 = ndimage.gaussian_filter(image, sigma1)
blurred2 = ndimage.gaussian_filter(image, sigma2)

# 计算 DoG 带通滤波器的输出图像
dog = blurred1 - blurred2

# 显示原始图像和 DoG 带通滤波器输出图像
fig, axes = plt.subplots(1, 2, figsize=(10, 5))
axes[0].imshow(image, cmap='gray')
axes[0].set_title('原始图像')
axes[0].axis('off')
axes[1].imshow(dog, cmap='gray')
axes[1].set_title('DoG 带通滤波器输出')
axes[1].axis('off')
plt.show()
```

在上述代码中，首先使用两个不同尺度的高斯模糊参数对图像进行高斯模糊操作，得到两个模糊图像。然后，通过计算两个模糊图像的差异，得到 DoG 带通滤波器的输出图像。代码执行后的效果如图 2-9 所示。

图 2-9　使用模块 ndimage 实现的效果

2.4.5 带阻滤波器

带阻滤波器（也称为陷波滤波器）是一种用于信号处理的滤波器，它可以抑制指定频率范围内的信号。与低通滤波器和高通滤波器不同，带阻滤波器允许某个频率范围之外的信号通过，而对于指定范围内的信号则进行抑制。

在 Python 程序中，可以使用 SciPy 库中的模块 signal 实现带阻滤波器。具体而言，可以使用函数 iirnotch() 来设计和应用带阻滤波器，该函数可以指定需要抑制的中心频率和带宽，从而创建一个带阻滤波器。例如，下面是一个使用模块 signal 实现带阻滤波器的例子。

实例 2-19：使用模块 signal 实现带阻滤波器

源码路径：daima\2\dlv.py

```python
import numpy as np
from scipy import signal
import matplotlib.pyplot as plt
# 设置中文字体
plt.rcParams["font.sans-serif"] = "SimHei"
plt.rcParams["axes.unicode_minus"] = False
# 生成示例信号
t = np.linspace(0, 1, 1000)
signal1 = np.sin(2 * np.pi * 10 * t)  # 10Hz 正弦信号
signal2 = np.sin(2 * np.pi * 60 * t)  # 60Hz 正弦信号
signal_noise = signal1 + signal2  # 合成的含噪声信号

# 设计带阻滤波器参数
fs = 1000  # 采样频率
f0 = 60  # 需要抑制的中心频率
Q = 30  # 带宽因子

# 创建带阻滤波器
b, a = signal.iirnotch(f0, Q, fs)

# 应用带阻滤波器
filtered_signal = signal.lfilter(b, a, signal_noise)

# 绘制原始信号和滤波后的信号
plt.figure(figsize=(10, 6))
plt.subplot(2, 1, 1)
plt.plot(t, signal_noise)
plt.title('原始信号')
plt.xlabel('时间')
plt.ylabel('幅值')

plt.subplot(2, 1, 2)
plt.plot(t, filtered_signal)
plt.title('滤波后的信号')
plt.xlabel('时间')
plt.ylabel('幅值')

plt.tight_layout()
plt.show()
```

在上述代码中，首先生成了一个含有 10Hz 和 60Hz 正弦信号的合成信号，并加入一些噪声。然后，使用函数 iirnotch()创建了一个带阻滤波器，指定了需要抑制的中心频率（60Hz）和带宽因子（Q）。最后，使用函数 lfilter()将带阻滤波器应用于信号，得到滤波后的结果。代码执行后的效果如图 2-10 所示。

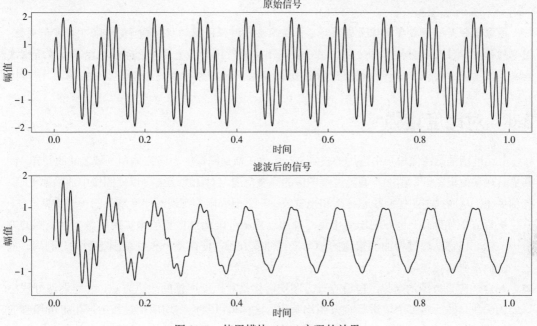

图 2-10　使用模块 signal 实现的效果

第 3 章 图像增强处理

图像增强是一种数字图像处理技术，旨在改善或增强图像的视觉质量、清晰度和可识别性。图像增强可以涉及多种技术和方法，用于提高图像的对比度、亮度、色彩鲜艳度、细节清晰度和噪声减少等方面。本章详细讲解使用 Python 语言实现图像增强的知识。

3.1 对比度增强

对比度增强是图像增强中常用的技术之一，旨在增强图像中不同物体和区域之间的差异，使其更具视觉效果和可识别性。通过调整图像的亮度范围，对比度增强可以使图像中的暗部更暗，亮部更亮，从而提高图像的动态范围和视觉效果。下面列出了实现对比度增强的一些常见方法。

- 直方图均衡化：直方图均衡化是一种通过重新分配图像像素值来扩展图像亮度范围的方法。该方法通过增加像素值的频率来拉伸直方图，使得整个亮度范围得到充分利用，从而增强对比度。
- 自适应直方图均衡化：自适应直方图均衡化是对比度增强的改进方法，它将图像分成小的块，然后对每个块进行直方图均衡化。这样可以避免在均衡化过程中产生过度增强的噪声，并且能够在局部区域中保留更多的细节。
- 对比度拉伸：对比度拉伸是通过线性映射将图像像素值映射到更广的亮度范围来增强对比度。它通过调整图像的最小和最大亮度值来实现，简单易懂，但可能会导致细节的丢失。
- 非线性对比度增强：非线性对比度增强方法根据像素值的分布特征对图像进行调整。常见的方法包括伽马校正、对数变换和指数变换等。

在实际应用中，可以根据图像的特点和需求选择合适的对比度增强方法。对比度增强可以应用于各种领域，如图像处理、计算机视觉、医学影像等，以提高图像的可视化效果、目标检测和图像分析的准确性。

3.1.1 直方图均衡化

直方图均衡化是一种常用的图像增强技术，通过重新分配图像的像素值来扩展图像的亮度范围，从而增强图像的对比度。下面是实现直方图均衡化的基本步骤：

（1）灰度化。如果原始图像是彩色图像，首先需要将其转换为灰度图像。这可以通过将彩色图像的 RGB 值转换为相应的灰度值来实现，常见的方法是取 RGB 值的平均值。

（2）计算直方图。对灰度图像进行直方图统计，即计算每个灰度级别（0~255）的像素数量。直方图可以表示为一个横轴为灰度级别，纵轴为对应灰度级别的像素数量的图表。

（3）计算累积直方图。通过对直方图进行累积求和，计算每个灰度级别的像素累积数量。这样可以得到一个新的直方图，其横轴表示灰度级别，纵轴表示对应灰度级别及以下的像素数量之和。

（4）灰度级别映射。将累积直方图进行线性映射，将原始图像中的每个像素值映射到新的灰度级别。映射公式为：

新像素值 = (累积直方图(原像素值) − 累积直方图最小值) / (像素总数 − 1) × 灰度级别最大值

其中，"累积直方图最小值"是累积直方图中的最小值；"像素总数"是图像的总像素数；"灰度级别最大值"是 255（8 位灰度图像的最大灰度级别）。

（5）应用灰度级别映射。用映射后的新像素值替换原始图像中的像素值。

（6）重复步骤（2）～（5）。对于图像的每个像素，重复步骤（2）～（5），直到对整个图像进行处理。

在 Python 程序中，可以通过如下两种方法实现直方图均衡化功能：

1．使有 OpenCV 库

使用 OpenCV 库中的函数 cv2.equalizeHist()可以对灰度图像进行直方图均衡化处理，它会自动计算直方图并进行均衡化，然后返回均衡化后的图像。例如，在下面的实例文件 ozhi.py 中，演示了使用 OpenCV 库实现直方图均衡化的过程。

实例 3-1：使用 OpenCV 库实现直方图均衡化

源码路径：daima\3\ozhi.py

```python
import cv2
import numpy as np

# 读取图像
image = cv2.imread('888.jpg', cv2.IMREAD_GRAYSCALE)

# 使用 OpenCV 的直方图均衡化函数
equalized_image = cv2.equalizeHist(image)

# 显示原始图像和均衡化后的图像
cv2.imshow("Original Image", image)
cv2.imshow("Equalized Image", equalized_image)
cv2.waitKey(0)
cv2.destroyAllWindows()
```

在上述代码中使用函数 cv2.equalizeHist()对灰度图像进行直方图均衡化，它会自动计算直方图并进行均衡化，然后返回均衡化后的图像。最后，我们使用 cv2.imshow()函数显示原始图像和均衡化后的图像。代码执行后的效果如图 3-1 所示。

图 3-1　使用 OpenCV 库实现的直方图均衡化效果

2. 使用 PIL 库

也可以使用 PIL 库中的函数 ImageOps.equalize()对灰度图像进行直方图均衡化，它会计算直方图并进行均衡化，然后返回均衡化后的图像。例如，在下面的实例文件 pzhi.py 中，演示了使用 PIL 库实现直方图均衡化的过程。

实例 3-2：使用 PIL 库实现直方图均衡化

源码路径：daima\3\pzhi.py

```python
from PIL import Image, ImageOps
import numpy as np

# 读取图像
image = Image.open('888.jpg').convert('L')

# 使用 PIL 的直方图均衡化函数
equalized_image = ImageOps.equalize(image)

# 显示原始图像和均衡化后的图像
image.show()
equalized_image.show()
```

注意：通过直方图均衡化，原始图像的像素值将根据其在直方图中的分布重新映射，使得图像中的亮度范围得到充分利用，增强了图像的对比度和视觉效果。这对于改善图像的可视化、目标检测和图像分析非常有用。然而，直方图均衡化也可能导致图像的局部细节损失，特别是当图像中的亮度变化较大时。在这种情况下，可以考虑应用自适应直方图均衡化等改进技术来保留更多的细节。

3.1.2 自适应直方图均衡化

自适应直方图均衡化（adaptive histogram equalization，AHE）是直方图均衡化的一种改进方法，它通过对图像的局部区域进行均衡化，以避免在全局均衡化中引入过度增强和噪声放大。在 Python 程序中，可以通过如下方法实现自适应直方图均衡化：

1. 使用 OpenCV 库

可以使用 OpenCV 库中的函数 cv2.createCLAHE()实现自适应直方图均衡化，例如下面的实例演示了这一用法。

实例 3-3：使用 OpenCV 库实现自适应直方图均衡化

源码路径：daima\3\zi.py

```python
import cv2
import numpy as np

# 读取图像
image = cv2.imread(888.jpg', cv2.IMREAD_GRAYSCALE)

# 使用 CLAHE（自适应直方图均衡化）算法
clahe = cv2.createCLAHE(clipLimit=2.0, tileGridSize=(8, 8))
equalized_image = clahe.apply(image)

# 显示原始图像和均衡化后的图像
```

```
cv2.imshow("Original Image", image)
cv2.imshow("Equalized Image", equalized_image)
cv2.waitKey(0)
cv2.destroyAllWindows()
```

在上述代码中，首先使用 OpenCV 库中的函数 cv2.createCLAHE()创建一个 CLAHE 对象（自适应直方图均衡化算法通过调整 clipLimit 参数来控制对比度增强的程度），参数 tileGridSize 定义了局部块的大小。然后，应用 CLAHE 对象到图像上，得到均衡化后的图像。代码执行后的效果如图 3-2 所示。

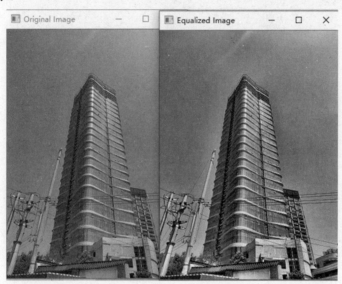

图 3-2　使用 OpenCV 实现的自适应直方图均衡化效果

2. 使用 scikit-image 库

可以使用 scikit-image 库中的函数 exposure.equalize_adapthist()实现自适应直方图均衡化。例如下面的实例演示了这一用法。

实例 3-4：使用库 scikit-image 实现自适应直方图均衡化

源码路径：daima\3\szi.py

```
from skimage import exposure
import cv2

# 读取图像
image = cv2.imread('888.jpg', cv2.IMREAD_GRAYSCALE)

# 使用自适应直方图均衡化算法
equalized_image = exposure.equalize_adapthist(image, clip_limit=0.03)

# 显示原始图像和均衡化后的图像
cv2.imshow("Original Image", image)
cv2.imshow("Equalized Image", equalized_image)
cv2.waitKey(0)
cv2.destroyAllWindows()
```

在上述代码中，首先使用库 scikit-image 中的 exposure.equalize_adapthist()函数进行自适应直方图均衡化，通过调整 t 参数 clip_limi 来控制对比度增强的程度。然后，应用该函数到图像上，得到均衡化后的图像。代码执行后的效果如图 3-3 所示。

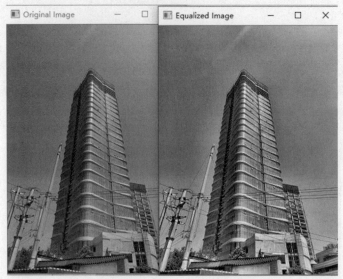

图 3-3　使用 scikit-image 实现的效果

注意：上面方法中的 CLAHE 对象和函数 equalize_adapthist() 都使用自适应均衡化算法，在局部区域上进行直方图均衡化，从而避免了全局均衡化中的过度增强和噪声放大。这些方法对于提高图像的对比度和视觉效果非常有用，可以应用于各种图像处理任务。

3.1.3　对比度拉伸

对比度拉伸（contrast stretching）是一种简单且常用的图像增强技术，通过线性映射来增加图像的对比度。它将图像的最小灰度值映射到较低的输出灰度级别，将最大灰度值映射到较高的输出灰度级别，从而扩展了图像的亮度范围，增强了图像的视觉效果。在 Python 程序中，可以通过如下方法实现对比度拉伸功能。

1. 使用 NumPy 和 OpenCV 库

例如，在下面的实例中，使用 NumPy 和 OpenCV 库实现了对比度拉伸。

实例 3-5：使用 NumPy 和 OpenCV 实现对比度拉伸

源码路径：daima\3\dui.py

```python
import cv2
import numpy as np

# 读取图像
image = cv2.imread('888.jpg', cv2.IMREAD_GRAYSCALE)

# 计算原始图像的最小和最大灰度值
min_value = np.min(image)
max_value = np.max(image)

# 执行对比度拉伸
stretched_image = (image - min_value) * (255.0 / (max_value - min_value))
```

```
# 将图像灰度值限制在 0~255 范围内
stretched_image = np.clip(stretched_image, 0, 255).astype(np.uint8)

# 显示原始图像和对比度拉伸后的图像
cv2.imshow("Original Image", image)
cv2.imshow("Stretched Image", stretched_image)
cv2.waitKey(0)
cv2.destroyAllWindows()
```

在上述代码中，首先使用函数 np.min()和 np.max()计算图像的最小和最大灰度值。然后，对图像进行线性映射，将像素值从最小到最大范围映射到 0~255 的范围。最后，使用函数 np.clip()将图像灰度值限制在 0~255，并将其转换为无符号 8 位整数类型。代码执行后的效果如图 3-4 所示。

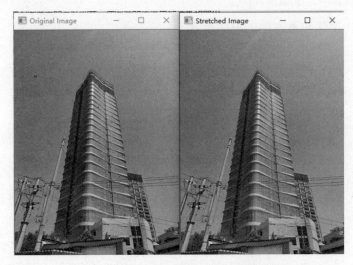

图 3-4　使用 NumPy 和 OpenCV 实现的效果

2. 使用 PIL 库

实例 3-6：使用 PIL 库实现对比度拉伸

例如，在下面的实例中，演示了使用 PIL 库实现对比度拉伸的过程。

源码路径：daima\3\pla.py

```
from PIL import Image
import numpy as np

# 读取图像
image = Image.open('888.jpg').convert('L')

# 执行对比度拉伸
min_value = np.min(image)
max_value = np.max(image)
stretched_image = (image - min_value) * (255.0 / (max_value - min_value))

# 将图像灰度值限制在 0~255 范围内
stretched_image = np.clip(stretched_image, 0, 255).astype(np.uint8)

# 显示原始图像和对比度拉伸后的图像
image.show()
```

```
Image.fromarray(stretched_image).show()
```

在上述代码中，首先使用函数 np.min()和 np.max()计算图像的最小和最大灰度值。然后，对图像进行线性映射，将像素值从最小到最大范围映射到 0~255 的范围。最后，使用函数 np.clip()将图像灰度值限制在 0~255，并将其转换为无符号 8 位整数类型。

3.1.4 非线性对比度增强

非线性对比度增强（nonlinear contrast enhancement）是一种图像增强技术，通过非线性的变换函数来调整图像的对比度，以增强图像的细节和视觉效果。与线性对比度增强方法不同，非线性对比度增强能够更好地处理图像的局部对比度变化。在 Python 程序中，可以通过如下几种方法实现非线性对比度增强功能：

1. 直方图均衡化和直方图匹配的组合

这种方法结合了直方图均衡化和直方图匹配的技术。首先，对图像进行直方图均衡化以增加全局对比度。然后，使用直方图匹配来进一步调整图像的对比度，使其更好地适应特定场景的对比度要求。例如，下面的实例演示了这一方法的实现过程。

实例 3-7：直方图均衡化和直方图匹配的非线性对比度增强

源码路径： daima\3\zhifei.py

```python
import cv2
import numpy as np

# 读取图像
image = cv2.imread('888.jpg', cv2.IMREAD_GRAYSCALE)

# 直方图均衡化
equalized_image = cv2.equalizeHist(image)

# 直方图匹配
target_histogram = cv2.calcHist([image], [0], None, [256], [0, 256])
target_histogram /= target_histogram.sum()
equalized_histogram = cv2.calcHist([equalized_image], [0], None, [256], [0, 256])
equalized_histogram /= equalized_histogram.sum()
mapping_function = np.cumsum(target_histogram) * 255
matched_image     =     np.interp(equalized_image.flatten(),     np.arange(256),
mapping_function).reshape(equalized_image.shape)

# 显示原始图像、直方图均衡化后的图像和非线性对比度增强后的图像
cv2.imshow("Original Image", image)
cv2.imshow("Equalized Image", equalized_image)
cv2.imshow("Enhanced Image", matched_image.astype(np.uint8))
cv2.waitKey(0)
cv2.destroyAllWindows()
```

在上述代码中，首先对图像进行直方图均衡化，然后计算原始图像和均衡化后图像的直方图，并将其归一化。接下来，根据均衡化后图像的直方图和目标直方图计算映射函数。最后，使用函数 np.interp()将均衡化后图像中的像素值映射到目标直方图的范围内，得到非线性对比度增强后的图像。代码执行后的效果如图 3-5 所示。

图 3-5　非线性对比度增强后的效果

2．对数变换

对数变换是一种常用的非线性对比度增强方法，它通过对像素值取对数的方式来调整图像的对比度。对数变换可以扩展低灰度级的细节，并压缩高灰度级的细节。例如，下面的实例演示了这一方法的实现过程。

实例 3-8：对数变换的非线性对比度增强

源码路径：daima\3\duifei.py

```python
import cv2
import numpy as np

# 读取图像
image = cv2.imread('888.jpg', cv2.IMREAD_GRAYSCALE)

# 对数变换
log_transformed_image = np.log1p(image)

# 将像素值归一化到 0～255 范围内
log_transformed_image = (255 * (log_transformed_image - np.min(log_transformed_image)) /
(np.max(log_transformed_image) - np.min(log_transformed_image))).astype (np.uint8)

# 显示原始图像和非线性对比度增强后的图像
cv2.imshow("Original Image", image)
cv2.imshow("Enhanced Image", log_transformed_image)
cv2.waitKey(0)
cv2.destroyAllWindows()
```

在上述代码中，使用 np.log1p()函数对图像进行对数变换。然后使用线性映射将像素值归一化到 0～255。最后，得到非线性对比度增强后的图像。

3.2　锐化

图像锐化（image sharpening）是一种图像处理技术，旨在增强图像的边缘和细节，使图像看起来更加清晰和鲜明。锐化技术通过增加图像的高频分量来突出边缘和细节，从而改善图像的视觉质量。图像锐化的主要目标是增强图像中的边缘信息。边缘是图像中灰度值变化较大的区域，通常表示物体的边界或纹理。通过突出边缘，图像锐化技术可以使图像看起来更加清晰，细节更加突出。

在现实应用中，有以下几种常见的图像锐化方法：

- 锐化滤波器：锐化滤波器是一种常用的图像锐化技术，通过增强高频分量来提高图像的清晰度。常见的锐化滤波器包括拉普拉斯滤波器和高通滤波器（如 sobel 滤波器和 prewitt 滤波器）。这些滤波器通过在图像上应用卷积操作来增强边缘。
- 高频强调滤波：高频强调滤波是一种通过增强图像的高频成分来实现图像锐化的技术。它基于图像锐化的原理，将原始图像与其低通滤波结果进行相减，从而突出高频细节。常用的高频强调滤波器包括 unsharp masking 和细节增强。
- 基于梯度的锐化：基于梯度的锐化方法利用图像的梯度信息来增强边缘。梯度是图像灰度变化最大的区域，通常与边缘相对应。通过计算图像的梯度，可以突出图像中的边缘信息。常见的基于梯度的锐化方法包括 sobel 算子和 canny 边缘检测算法。

Python 中有多种库可以实现图像锐化，例如 OpenCV 和 PIL。使用这些库，可以方便地应用各种图像锐化技术来增强图像的清晰度和细节。具体使用哪种方法实现图像锐化功能，取决于所选择的库和方法。

3.2.1 锐化滤波器

锐化滤波器（sharpening filter）是一种常用的图像处理技术，用于增强图像的边缘和细节，以提高图像的清晰度和视觉效果。锐化滤波器通过增强图像中的高频分量来突出边缘，从而使图像看起来更加清晰和鲜明。在图像处理中，常见的锐化滤波器包括拉普拉斯滤波器和高通滤波器，如 sobel 滤波器和 prewitt 滤波器。

- 拉普拉斯滤波器是一种常用的锐化滤波器，用于增强图像的边缘信息。它通过对图像进行二阶微分来检测边缘，然后将检测到的边缘添加回原始图像以增强边缘。在实现时，常用的拉普拉斯滤波器有 3×3 和 5×5 两种核。
- sobel 滤波器是一种常用的高通滤波器，用于检测图像中的边缘。通过计算图像的梯度来确定像素值的变化情况，并突出边缘。sobel 滤波器分为水平和垂直两个方向，可以分别检测图像中的水平和垂直边缘。通过对这两个方向的边缘进行组合，可以得到更全面的边缘检测结果。

Python 中，可以使用 OpenCV 库来实现锐化滤波器。

1. 使用拉普拉斯滤波器实现图像锐化

例如，在下面的实例中，演示了使用拉普拉斯滤波器实现图像锐化的过程。

实例 3-9：使用拉普拉斯滤波器实现图像锐化

源码路径：daima\3\purui.py

```
import cv2
import numpy as np

# 读取图像
image = cv2.imread('888.jpg', cv2.IMREAD_GRAYSCALE)
```

```python
# 定义拉普拉斯滤波器
laplacian_kernel = np.array([[0, 1, 0],
                             [1, -4, 1],
                             [0, 1, 0]], dtype=np.float32)

# 对图像应用拉普拉斯滤波器
sharpened_image = cv2.filter2D(image, -1, laplacian_kernel)

# 将像素值归一化到 0～255 范围内
sharpened_image = (255 * (sharpened_image - np.min(sharpened_image)) / (np.max(sharpened_image) - np.min(sharpened_image))).astype(np.uint8)

# 显示原始图像和锐化后的图像
cv2.imshow("Original Image", image)
cv2.imshow("Sharpened Image", sharpened_image)
cv2.waitKey(0)
cv2.destroyAllWindows()
```

在上述代码中，首先使用函数 cv2.imread()读取图像，并将其转换为灰度图像。然后定义一个 3×3 的拉普拉斯滤波器作为卷积核。通过使用函数 cv2.filter2D()对图像应用拉普拉斯滤波器，可以得到锐化后的图像。最后，将锐化后的图像像素值归一化到 0～255 的范围内，并显示原始图像和锐化后的图像。注意，在本实例中将锐化后的图像的像素值归一化到 0～255 的范围内，并将其转换为 np.uint8 类型。但是，如果原始图像的像素值范围本身已经是 0～255，那么进行归一化可能会导致所有的像素值都变为 0，从而得到一片黑色的图像。此时可以尝试去掉像素值归一化的步骤，直接使用滤波后的图像进行显示。以下是修改后的代码：

```python
import cv2
import numpy as np

# 读取图像
image = cv2.imread('888.jpg', cv2.IMREAD_GRAYSCALE)

# 定义拉普拉斯滤波器
laplacian_kernel = np.array([[0, 1, 0],
                             [1, -4, 1],
                             [0, 1, 0]], dtype=np.float32)

# 对图像应用拉普拉斯滤波器
sharpened_image = cv2.filter2D(image, -1, laplacian_kernel)

# 显示原始图像和锐化后的图像
cv2.imshow("Original Image", image)
cv2.imshow("Sharpened Image", sharpened_image)
cv2.waitKey(0)
cv2.destroyAllWindows()
```

代码执行后的效果如图 3-6 所示。

2. 使用 sobel 滤波器实现图像锐化

例如，在下面的实例中，使用 sobel 滤波器计算图像的梯度，并通过加权合并水平和垂直梯度来实现图像的锐化效果。我们可以尝试在代码中修改参数，如调整权重值、修改滤波器的大小等，以获得不同的锐化效果。

图 3-6 使用拉普拉斯滤波器实现的效果

实例 3-10：使用 sobel 滤波器实现图像锐化
源码路径：daima\3\srui.py

```python
import cv2

# 读取图像
image = cv2.imread('image.jpg', cv2.IMREAD_GRAYSCALE)

# 计算水平方向和垂直方向的梯度
gradient_x = cv2.Sobel(image, cv2.CV_64F, 1, 0, ksize=3)
gradient_y = cv2.Sobel(image, cv2.CV_64F, 0, 1, ksize=3)

# 合并梯度
sharpened_image = cv2.addWeighted(gradient_x, 0.5, gradient_y, 0.5, 0)

# 显示原始图像和锐化后的图像
cv2.imshow("Original Image", image)
cv2.imshow("Sharpened Image", sharpened_image)
cv2.waitKey(0)
cv2.destroyAllWindows()
```

对上述代码的具体说明如下：

（1）导入库：首先导入了 OpenCV 库，它提供了图像处理和计算机视觉相关的功能。

（2）读取图像：使用 cv2.imread() 函数读取名为 image.jpg 的图像，并通过 cv2.IMREAD_GRAYSCALE 参数将其转换为灰度图像。

（3）计算梯度：通过使用 sobel 滤波器计算图像的水平和垂直方向的梯度。cv2.Sobel() 函数接受几个参数：第一个参数是输入图像，第二个参数是输出图像的数据类型（这里使用 cv2.CV_64F 表示 64 位浮点数），第三个和第四个参数分别是水平和垂直方向的导数阶数，最后一个参数是滤波器的大小（这里使用 3×3 的滤波器）。

（4）合并梯度：通过使用 cv2.addWeighted() 函数将水平和垂直方向的梯度进行加权合并。这里将两个梯度图像按照相等的权重（0.5）进行加权合并，并将结果存储在 sharpened_image 变量中。

（5）显示图像：使用 cv2.imshow() 函数显示原始图像和锐化后的图像。cv2.waitKey(0) 函数等待用户按下任意键来关闭显示窗口。最后，使用 cv2.destroyAllWindows() 函数关闭所有的显示窗口。

3.2.2 高频强调滤波

高频强调滤波（high-frequency emphasis filtering）是一种图像增强技术，用于增强图像中的高频细节和边缘。高频强调滤波通过突出图像的高频分量，使细节更加清晰和明显。Python语言实现高频强调滤波的方法主要包括以下几种。

1．理想高通滤波器

具体实现步骤如下：

（1）使用傅里叶变换将图像转换到频域；

（2）在频域中，使用理想高通滤波器将低频分量设置为零，保留高频分量；

（3）将频域图像通过傅里叶反变换转换回空域图像。

例如，下面是一个使用理想高通滤波器实现高频强调滤波的例子。

实例 3-11：使用理想高通滤波器实现高频强调滤波

源码路径： daima\3\gaotong.py

```python
import cv2
import numpy as np

def ideal_highpass_filter(image, cutoff):
    # 将图像转换到频域
    dft = cv2.dft(np.float32(image), flags=cv2.DFT_COMPLEX_OUTPUT)
    dft_shift = np.fft.fftshift(dft)

    # 构建理想高通滤波器
    rows, cols = image.shape
    center_row, center_col = rows // 2, cols // 2
    distance = np.sqrt((np.arange(rows)[:, np.newaxis] - center_row) ** 2 +
                       (np.arange(cols) - center_col) ** 2)
    highpass = np.ones_like(image)
    highpass[distance <= cutoff] = 0

    # 应用滤波器
    dft_shift_filtered = dft_shift * highpass[:, :, np.newaxis]

    # 将频域图像转换回空域
    dft_filtered_shifted = np.fft.ifftshift(dft_shift_filtered)
    filtered_image = cv2.idft(dft_filtered_shifted)
    filtered_image = cv2.magnitude(filtered_image[:, :, 0], filtered_image[:, :, 1])

    # 保持原始图像的亮度范围
    filtered_image = filtered_image * 255 / np.max(filtered_image)

    return filtered_image.astype(np.uint8)

# 读取图像
image = cv2.imread('888.jpg', cv2.IMREAD_GRAYSCALE)

# 应用理想高通滤波器
enhanced_image = ideal_highpass_filter(image, cutoff=50)

# 显示原始图像和增强后的图像
```

```
cv2.imshow("Original Image", image)
cv2.imshow("Enhanced Image", enhanced_image)
cv2.waitKey(0)
cv2.destroyAllWindows()
```

在上述代码中，使用函数 ideal_highpass_filter()实现了理想高通滤波器。先将图像转换到频域，再构建理想高通滤波器，将低频分量设为零。接下来，将滤波器应用于频域图像，并将结果转换回空域图像。然后，使用函数 cv2.imread()读取图像，并将其转换为灰度图像。调用函数 ideal_highpass_filter()，传入图像和截止频率 cutoff 进行理想高通滤波。最后，使用函数 cv2.imshow()显示原始图像和增强后的图像。我们也可以尝试调整截止频率的值 cutoff，以获得不同的高频强调效果。较低的截止频率会保留更多的低频分量，而较高的截止频率会突出高频细节。代码执行后的效果如图 3-7 所示。

图 3-7 使用理想高通滤波器实现的效果

2. 巴特沃斯高通滤波器

巴特沃斯高通滤波器提供了平滑的过渡区域，避免了理想滤波器的陡峭截止。例如，下面是一个使用巴特沃斯高通滤波器实现高频强调滤波的例子。

实例 3-12：使用巴特沃斯高通滤波器实现高频强调滤波

源码路径：daima\3\bate.py

```python
import cv2
import numpy as np

def high_frequency_emphasis(image, alpha, cutoff):
    # 将图像转换到频域
    dft = cv2.dft(np.float32(image), flags=cv2.DFT_COMPLEX_OUTPUT)
    dft_shift = np.fft.fftshift(dft)

    # 构建巴特沃斯高通滤波器
    rows, cols = image.shape
    center_row, center_col = rows // 2, cols // 2
    distance = np.sqrt((np.arange(rows)[:, np.newaxis] - center_row) ** 2 +
                       (np.arange(cols) - center_col) ** 2)
    highpass = 1 / (1 + (cutoff / distance) ** (2 * alpha))

    # 应用滤波器
    dft_shift_filtered = dft_shift * highpass[:, :, np.newaxis]

    # 将频域图像转换回空域
    dft_filtered_shifted = np.fft.ifftshift(dft_shift_filtered)
    filtered_image = cv2.idft(dft_filtered_shifted)
    filtered_image = cv2.magnitude(filtered_image[:, :, 0], filtered_image[:, :, 1])

    # 保持原始图像的亮度范围
```

```
        filtered_image = filtered_image * 255 / np.max(filtered_image)

        return filtered_image.astype(np.uint8)

# 读取图像
image = cv2.imread('888.jpg', cv2.IMREAD_GRAYSCALE)

# 应用高频强调滤波
enhanced_image = high_frequency_emphasis(image, alpha=2, cutoff=50)

# 显示原始图像和增强后的图像
cv2.imshow("Original Image", image)
cv2.imshow("Enhanced Image", enhanced_image)
cv2.waitKey(0)
cv2.destroyAllWindows()
```

在上述代码中,首先定义了一个名为 high_frequency_emphasis()的函数,它以输入图像、强调参数 alpha 和截止频率 cutoff 作为输入。该函数实现了巴特沃斯高通滤波器的过程,包括傅里叶变换、滤波器构建、滤波和逆傅里叶变换等步骤。然后,使用函数 cv2.imread()读取图像,并将其转换为灰度图像。接下来,调用函数 high_frequency_emphasis(),传入图像和参数进行高频强调滤波。最后,使用函数 cv2.imshow()显示原始图像和增强后的图像。也可以尝试调整参数 alpha 和参数 cutoff 的值,以获得不同的高频强调效果。代码执行后的效果如图 3-8 所示。

3. 带通滤波器

带通滤波器通过将低频和高频分量同时保留,滤波掉中间频率的分量,从而突出高频细节。例如,下面是一个使用带通滤波器实现高频强调滤波的例子。

图 3-8 使用巴特沃斯高通滤波器实现的效果

实例 3-13:使用带通滤波器实现高频强调滤波

源码路径:daima\3\dai.py

```
import cv2
import numpy as np

def bandpass_filter(image, low_cutoff, high_cutoff):
    # 将图像转换到频域
    dft = cv2.dft(np.float32(image), flags=cv2.DFT_COMPLEX_OUTPUT)
    dft_shift = np.fft.fftshift(dft)

    # 构建带通滤波器
    rows, cols = image.shape
    center_row, center_col = rows // 2, cols // 2
    distance = np.sqrt((np.arange(rows)[:, np.newaxis] - center_row) ** 2 +
                       (np.arange(cols) - center_col) ** 2)
```

```python
    bandpass = np.zeros_like(image)
    bandpass[(distance >= low_cutoff) & (distance <= high_cutoff)] = 1

    # 应用滤波器
    dft_shift_filtered = dft_shift * bandpass[:, :, np.newaxis]

    # 将频域图像转换回空域
    dft_filtered_shifted = np.fft.ifftshift(dft_shift_filtered)
    filtered_image = cv2.idft(dft_filtered_shifted)
    filtered_image = cv2.magnitude(filtered_image[:, :, 0], filtered_image[:, :, 1])

    # 保持原始图像的亮度范围
    filtered_image = filtered_image * 255 / np.max(filtered_image)

    return filtered_image.astype(np.uint8)

# 读取图像
image = cv2.imread('image.jpg', cv2.IMREAD_GRAYSCALE)

# 应用带通滤波器
enhanced_image = bandpass_filter(image, low_cutoff=20, high_cutoff=80)

# 显示原始图像和增强后的图像
cv2.imshow("Original Image", image)
cv2.imshow("Enhanced Image", enhanced_image)
cv2.waitKey(0)
cv2.destroyAllWindows()
```

对上述代码的具体说明如下：

（1）函数 bandpass_filter()实现了带通滤波器的过程。首先将图像转换到频域，然后构建一个二值滤波器，其中在低截止频率和高截止频率之间的频率范围内取值为 1，其余为 0。接下来，将滤波器应用于频域图像，并将结果转换回空域图像。

（2）使用函数 cv2.imread()读取图像，并将其转换为灰度图像。调用函数 bandpass_filter()，传入图像和截止频率范围进行带通滤波。

（3）使用函数 cv2.imshow()显示原始图像和增强后的图像。可以尝试调整低截止频率和高截止频率的值，以获得不同的高频强调效果。较小的截止频率会保留更多的低频分量，较大的截止频率会突出高频细节。

3.2.3 基于梯度的锐化

基于梯度的锐化是一种图像增强技术，通过突出图像中的边缘和细节来增强图像的清晰度和锐度。该方法基于图像的梯度信息，利用梯度的变化来增强图像的边缘。

在基于梯度的锐化方法中，常用的操作包括边缘检测和梯度增强。边缘检测算法可用于提取图像中的边缘信息，如 sobel、prewitt 和 canny 等。梯度增强算法可增强图像中的边缘，如拉普拉斯算子、高频增强滤波器等。

1. sobel 算子

sobel 算子是一种常用的边缘检测算子，用于在图像中寻找边缘的位置和方向。它是基于图像中的灰度变化率来进行边缘检测的。sobel 算子分别计算了图像在水平和垂直方向上的一阶导数。通过计算这两个方向上的梯度，可以获取图像中的边缘信息。sobel 算子是基于离散卷积的操作，它在图像的每个像素上应用一个 3×3 的卷积核。sobel 算子的卷积核如下：

```
        | -1    0    1 |
Sx  =   | -2    0    2 |
        | -1    0    1 |


        | -1   -2   -1 |
Sy  =   |  0    0    0 |
        |  1    2    1 |
```

其中，Sx 代表水平方向的 sobel 算子，Sy 代表垂直方向的 sobel 算子。

sobel 算子的运算过程如下：

（1）对图像进行灰度转换（如果图像不是灰度图像）。
（2）分别使用 Sx 和 Sy 卷积核对图像进行卷积操作，得到水平和垂直方向上的梯度值。
（3）计算每个像素的梯度幅值和方向。梯度幅值：sqrt(Sx^2 + Sy^2)；梯度方向：atan2(Sy, Sx)。
（4）对梯度幅值进行阈值处理，以提取边缘。
（5）根据需要，可将提取的边缘绘制在图像上或进行其他后续处理。

sobel 算子可用于边缘检测、图像锐化、特征提取等图像处理任务。它的优点是简单高效，并且对噪声具有一定的抑制作用。例如，下面的实例演示了使用 sobel 算子实现基于梯度的锐化的过程。

实例 3-14：使用 sobel 算子实现基于梯度的锐化
源码路径：daima\3\sobel.py

```python
import cv2
import numpy as np

def sobel_sharpen(image):
    # 计算水平和垂直方向的梯度
    gradient_x = cv2.Sobel(image, cv2.CV_64F, 1, 0, ksize=3)
    gradient_y = cv2.Sobel(image, cv2.CV_64F, 0, 1, ksize=3)

    # 取绝对值并合并梯度
    gradient_x = cv2.convertScaleAbs(gradient_x)
    gradient_y = cv2.convertScaleAbs(gradient_y)
    gradient = cv2.addWeighted(gradient_x, 0.5, gradient_y, 0.5, 0)

    # 对原始图像和梯度图像进行加权叠加
    sharpened_image = cv2.addWeighted(image, 0.5, gradient, 0.5, 0)

    return sharpened_image

# 读取图像
image = cv2.imread('888.jpg', cv2.IMREAD_GRAYSCALE)

# 应用sobel算子进行锐化
sharpened_image = sobel_sharpen(image)

# 显示原始图像和锐化后的图像
cv2.imshow("Original Image", image)
cv2.imshow("Sharpened Image (Sobel)", sharpened_image)
cv2.waitKey(0)
cv2.destroyAllWindows()
```

对上述代码的具体说明如下：

（1）定义函数 sobel_sharpen()，该函数以一个灰度图像作为输入，并返回锐化后的图像。

（2）在函数 sobel_sharpen()内部，使用函数 cv2.Sobel()计算了图像在水平和垂直方向上的梯度。其中，gradient_x 表示水平方向的梯度，gradient_y 表示垂直方向的梯度。cv2.CV_64F 指定了输出图像的数据类型为 64 位浮点数。

（3）使用 cv2.convertScaleAbs()函数对梯度图像进行绝对值转换，并将结果存储在 gradient_x 和 gradient_y 中。这一步是为了保证梯度图像的数值范围在 0～255。

（4）使用 cv2.addWeighted()函数将水平和垂直方向上的梯度图像进行加权叠加，得到合并后的梯度图像。这里设置了相同的权重 0.5，表示对两个梯度图像进行平均。

（5）使用 cv2.addWeighted()函数将原始图像和合并后的梯度图像进行加权叠加，得到最终的锐化图像。这里也设置了相同的权重 0.5，表示对两个图像进行平均。

（6）在主程序中，使用 cv2.imread()函数读取了一张灰度图像（假设文件名为'image.jpg'）。

（7）调用 sobel_sharpen()函数对图像进行锐化，得到锐化后的图像 sharpened_image。

（8）使用 cv2.imshow()函数显示原始图像和锐化后的图像。

（9）使用 cv2.waitKey(0)函数等待用户按下任意键后关闭显示窗口。

（10）最后，使用 cv2.destroyAllWindows()函数关闭所有显示窗口。

总体来说，该代码通过计算图像的梯度，并将梯度图像与原始图像进行加权叠加，实现了对图像的锐化处理。

2．laplacian 算子

laplacian 算子是一种常用的边缘检测算子，用于在图像中寻找边缘的位置和方向。它基于图像中的二阶导数，可以更好地捕捉到图像中的高频变化。laplacian 算子对图像进行了二次微分运算，从而可以检测出图像中的局部变化和突变。它在图像的每个像素点上应用了一个拉普拉斯模板（通常是 3×3 的模板），计算图像中的像素值与其周围像素值之间的差异。

laplacian 算子的卷积核如下：

```
        | 0  1  0 |
  L =   | 1 -4  1 |
        | 0  1  0 |
```

其中，L 代表 laplacian 算子。例如，下面的实例演示了使用 laplacian 算子实现基于梯度的锐化的过程。

实例 3-15：使用 laplacian 算子实现基于梯度的锐化

源码路径： daima\3\lap.py

```python
import cv2

def laplacian_sharpen(image):
    # 应用拉普拉斯算子进行锐化
    laplacian = cv2.Laplacian(image, cv2.CV_64F)
    sharpened_image = cv2.convertScaleAbs(image - laplacian)

    return sharpened_image

# 读取图像
image = cv2.imread('888.jpg', cv2.IMREAD_GRAYSCALE)
```

```
# 应用Laplacian算子进行锐化
sharpened_image = laplacian_sharpen(image)

# 显示原始图像和锐化后的图像
cv2.imshow("Original Image", image)
cv2.imshow("Sharpened Image (Laplacian)", sharpened_image)
cv2.waitKey(0)
cv2.destroyAllWindows()
```

对上述代码的具体说明如下:

(1) 定义了 laplacian_sharpen()函数,该函数接受一个灰度图像作为输入,并返回锐化后的图像。

(2) 在 laplacian_sharpen()函数内部,使用 cv2.Laplacian 函数对图像应用了 laplacian 算子。其中,cv2.CV_64F 指定了输出图像的数据类型为 64 位浮点数。

(3) 使用 cv2.convertScaleAbs()函数将 laplacian 算子的结果取绝对值并转换为无符号 8 位整数,得到锐化后的图像 sharpened_image。

(4) 在主程序中,使用 cv2.imread()函数读取了一张灰度图像(假设文件名为'image.jpg')。

(5) 调用 laplacian_sharpen()函数对图像进行锐化,得到锐化后的图像 sharpened_image。

(6) 使用 cv2.imshow()函数显示原始图像和锐化后的图像。

(7) 使用 cv2.waitKey(0)函数等待用户按下任意键后关闭显示窗口。

(8) 最后,使用 cv2.destroyAllWindows()函数关闭所有显示窗口。

总体来说,上述代码使用 Laplacian 算子对图像进行了二次微分操作,通过计算图像的二阶导数来实现图像的锐化处理。

3. 高频增强滤波器

高频增强滤波器是一种用于增强图像高频信息的滤波器。在图像处理中,高频成分通常对应着图像的细节和边缘信息。通过增强高频成分,可以使图像的细节更加清晰和突出。高频增强滤波器的原理是通过减小图像中的低频成分,从而增强高频成分。

在 Python 中,可以使用 NumPy 和 OpenCV 等库来实现高频增强滤波器。具体的实现方法可能会根据所选择的滤波器类型而有所不同,如使用巴特沃斯滤波器、高斯滤波器或理想滤波器等。这些滤波器通常需要通过设置参数,如截止频率、阶数或滤波器大小等来调整滤波器的性能。例如,下面的实例代码演示了如何使用高斯滤波器实现高频增强滤波器的过程。

实例 3-16:使用高斯滤波器实现高频增强滤波器
源码路径: daima\3\gao.py

```
import cv2
import numpy as np

def high_frequency_enhancement(image, sigma):
    # 将图像转换为灰度图像
    gray_image = cv2.cvtColor(image, cv2.COLOR_BGR2GRAY)
    
    # 使用高斯滤波平滑图像
    blurred_image = cv2.GaussianBlur(gray_image, (0, 0), sigma)
    
    # 计算图像的细节部分
    detail_image = gray_image - blurred_image
```

```python
    # 对细节部分进行增强
    enhanced_image = gray_image + detail_image

    return enhanced_image

# 读取图像
image = cv2.imread('888.jpg')

# 设置高斯滤波器的标准差
sigma = 3.0

# 应用高频增强滤波器
enhanced_image = high_frequency_enhancement(image, sigma)

# 显示原始图像和增强后的图像
cv2.imshow('Original Image', image)
cv2.imshow('Enhanced Image', enhanced_image)
cv2.waitKey(0)
cv2.destroyAllWindows()
```

在上述代码中，使用高斯滤波器对图像进行平滑处理，并计算出图像的细节部分。然后，将细节部分加回到原始图像中，得到增强了高频信息的图像。通过调整高斯滤波器的标准差参数，可以控制平滑的程度和高频增强的效果。代码执行后的效果如图 3-9 所示。

图 3-9　使用高斯滤波器实现的效果

3.3　噪声减少

在图像处理领域，通过降低图像中的噪声水平可以改善图像的视觉质量。在现实中，常用的噪声减少方法包括平滑滤波和去噪算法。

3.3.1　均值滤波器

均值滤波器是一种简单且常用的噪声减少技术。它通过在图像中的每个像素周围取一个固定大小的窗口，计算窗口中所有像素的平均值，并将该平均值赋给中心像素。这种方法对于高斯噪声和均匀噪声的去除效果较好，但可能会导致图像的模糊。例如，在下面的实例中，演示了使用均值滤波器实现图像噪声减少的过程。

实例 3-17：使用均值滤波器实现图像噪声减少

源码路径：daima\3\junjian.py

```python
import cv2
import numpy as np

# 读取图像
image = cv2.imread('888.jpg', cv2.IMREAD_COLOR)

# 将图像转换为灰度图像
gray_image = cv2.cvtColor(image, cv2.COLOR_BGR2GRAY)

# 定义均值滤波器的窗口大小
kernel_size = 5

# 使用均值滤波器进行滤波
filtered_image = cv2.blur(gray_image, (kernel_size, kernel_size))

# 显示原始图像和滤波后的图像
cv2.imshow('Original Image', gray_image)
cv2.imshow('Filtered Image', filtered_image)
cv2.waitKey(0)
cv2.destroyAllWindows()
```

在上述代码中，首先使用 OpenCV 库读取一张彩色图像，并将其转换为灰度图像。然后，定义了均值滤波器的窗口大小（这里是 5×5）。最后，使用 cv2.blur() 函数应用均值滤波器进行滤波，并将滤波后的图像显示出来。通过调整 kernel_size 值控制滤波器的窗口大小，从而影响滤波的效果。较大的窗口可以更有效地平滑图像，但可能会导致细节的丢失。代码执行后的效果如图 3-10 所示。

图 3-10 使用均值滤波器实现的效果

3.3.2 中值滤波器

中值滤波器是一种非线性滤波器，对于脉冲噪声的去除效果较好。它的原理是在窗口中取所有像素的中值，并将中值赋给中心像素。中值滤波器能够有效去除离群值，但可能会导致图像细节的损失。例如，下面是一个使用中值滤波器实现图像噪声减少的例子。

实例 3-18：使用中值滤波器实现图像噪声减少

源码路径：daima\3\zhong.py

```python
import cv2
import numpy as np
```

```python
# 读取图像
image = cv2.imread('image.jpg', cv2.IMREAD_COLOR)

# 将图像转换为灰度图像
gray_image = cv2.cvtColor(image, cv2.COLOR_BGR2GRAY)

# 定义中值滤波器的窗口大小
kernel_size = 5

# 使用中值滤波器进行滤波
filtered_image = cv2.medianBlur(gray_image, kernel_size)

# 显示原始图像和滤波后的图像
cv2.imshow('Original Image', gray_image)
cv2.imshow('Filtered Image', filtered_image)
cv2.waitKey(0)
cv2.destroyAllWindows()
```

在上述代码中，OpenCV 库读取一张彩色图像，并将其转换为灰度图像。然后，定义了中值滤波器的窗口大小（这里是 5×5）。最后，使用 cv2.medianBlur()函数应用中值滤波器进行滤波，并将滤波后的图像显示出来。通过调整 kernel_size 的值控制滤波器的窗口大小，从而影响滤波的效果。脉冲性噪声，但可能会导致细节的损失。

3.3.3 高斯滤波器

高斯滤波器是一种线性滤波器，它基于高斯函数对像素进行加权平均。它能够在滤波过程中保留边缘信息，并对高斯噪声有较好的去除效果。高斯滤波器的滤波窗口大小和标准差可以调整以平衡去噪效果和保留图像细节之间的权衡。例如，下面是一个使用高斯滤波器实现图像噪声减少的例子。

实例 3-19：使用高斯滤波器实现图像噪声减少

源码路径：daima\3\gaolv.py

```python
import cv2
import numpy as np

# 读取图像
image = cv2.imread('image.jpg', cv2.IMREAD_COLOR)

# 将图像转换为灰度图像
gray_image = cv2.cvtColor(image, cv2.COLOR_BGR2GRAY)

# 定义高斯滤波器的窗口大小和标准差
kernel_size = 5
sigma = 1.5

# 使用高斯滤波器进行滤波
filtered_image = cv2.GaussianBlur(gray_image, (kernel_size, kernel_size), sigma)

# 显示原始图像和滤波后的图像
cv2.imshow('Original Image', gray_image)
cv2.imshow('Filtered Image', filtered_image)
cv2.waitKey(0)
```

```
cv2.destroyAllWindows()
```

在上述代码中，首先使用 OpenCV 库读取一张彩色图像，并将其转换为灰度图像。然后，定义了高斯滤波器的窗口大小（在这里是 5×5）和标准差（sigma，控制滤波器的平滑程度）。最后，我们使用 cv2.GaussianBlur()函数应用高斯滤波器进行滤波，并将滤波后的图像显示出来。通过调整 kernel_size 的大小和 sigma 的值，可以控制滤波器的窗口大小和平滑程度，从而影响滤波的效果。

3.3.4 双边滤波器

双边滤波器是一种非线性滤波器，结合了空间域和像素值域的相似性。它通过考虑像素之间的空间距离和灰度值差异来进行滤波。双边滤波器能够保留边缘细节，并对噪声进行有效的抑制。例如，下面是一个使用双边滤波器实现图像噪声减少的例子。

实例 3-20：使用双边滤波器实现图像噪声减少

源码路径：daima\3\shuang.py

```
import cv2
import numpy as np

# 读取图像
image = cv2.imread('image.jpg', cv2.IMREAD_COLOR)

# 定义双边滤波器的参数
d = 15  # 邻域直径
sigma_color = 75  # 颜色空间标准差
sigma_space = 75  # 坐标空间标准差

# 使用双边滤波器进行滤波
filtered_image = cv2.bilateralFilter(image, d, sigma_color, sigma_space)

# 显示原始图像和滤波后的图像
cv2.imshow('Original Image', image)
cv2.imshow('Filtered Image', filtered_image)
cv2.waitKey(0)
cv2.destroyAllWindows()
```

在上述代码中，首先使用 OpenCV 库读取一张彩色图像。然后，定义了双边滤波器的参数，包括邻域直径 d、颜色空间标准差 sigma_color 和坐标空间标准差 sigma_space。最后，使用 cv2.bilateralFilter()函数应用双边滤波器进行滤波，并将滤波后的图像显示出来。双边滤波器在滤波时考虑了像素之间的空间距离和灰度值差异，因此能够保留边缘细节，并对噪声进行有效地抑制。

3.3.5 小波降噪

小波降噪是一种基于小波变换的噪声减少技术。小波变换可以将信号分解成不同尺度的频带，噪声通常集中在高频带中。通过对小波系数进行阈值处理，可以将噪声系数设置为零或减小其幅值，然后再进行小波逆变换，恢复图像。小波降噪能够在去噪的同时保留图像的细节和边缘信息。例如，下面是一个使用小波降噪实现图像噪声减少的例子。

实例 3-21：使用小波降噪实现图像噪声减少

源码路径：daima\3\xiao.py

```python
import cv2
import numpy as np
import pywt

# 读取图像
image = cv2.imread('image.jpg', cv2.IMREAD_GRAYSCALE)

# 小波降噪参数
wavelet = 'db1'  # 选取小波函数
level = 3  # 分解的层数

# 对图像进行小波分解
coeffs = pywt.wavedec2(image, wavelet, level=level)

# 降噪处理
coeffs_threshold = list(coeffs)
threshold = 10  # 设定阈值

for i in range(1, len(coeffs_threshold)):
    coeffs_threshold[i] = tuple(
        pywt.threshold(c, threshold) for c in coeffs_threshold[i]
    )

# 对图像进行小波重构
image_denoised = pywt.waverec2(coeffs_threshold, wavelet)

# 将像素值限制在 0～255 之间
image_denoised = np.clip(image_denoised, 0, 255)

# 将图像转换为 uint8 类型
image_denoised = image_denoised.astype(np.uint8)

# 显示原始图像和降噪后的图像
cv2.imshow('Original Image', image)
cv2.imshow('Denoised Image', image_denoised)
cv2.waitKey(0)
cv2.destroyAllWindows()
```

在上述代码中，首先使用 OpenCV 库读取一张灰度图像。然后，选择了小波函数（这里选择了 Daubechies 1 小波函数）和分解的层数（这里选择了 3 级）。使用 pywt.wavedec2() 函数对图像进行小波分解，得到系数数组。接下来，我们对系数进行阈值处理，将小于阈值的系数置为零。最后，使用 pywt.waverec2() 函数对处理后的系数进行小波重构，得到降噪后的图像。通过调整参数 wavelet 和参数 level 可以选择不同的小波函数和分解层数，从而影响降噪的效果。调整参数 threshold 可以控制阈值的大小，进一步调节降噪效果。

3.4 色彩平衡

色彩平衡是指调整图像的色调、饱和度和亮度，以获得更准确的颜色表示或实现特定的视觉效果。要通过色彩平衡来实现图像增强，可以使用直方图均衡化技术。直方图均衡化可以增强图像的对比度和动态范围，使图像的颜色分布更加均衡。

3.4.1 白平衡

白平衡是一种色彩平衡技术，用于校正图像中的色温偏移，以使白色物体在不同光照条件下呈现出相似的色彩。它的目的是消除图像中的色偏，使得白色物体看起来真实且中性。白平衡校正可以根据光源的颜色温度来调整图像的色调，从而使图像的整体色彩看起来更加平衡和自然。

一种常用的白平衡算法是基于灰度世界假设。该假设认为在自然光照条件下，整个场景的平均反射率在 RGB 颜色通道上是相等的。因此，通过计算每个颜色通道的平均值，并将其视为灰度世界中的中性灰色，然后根据这个中性灰色来调整图像的颜色，达到色彩平衡的效果。例如，下面是一个使用基于灰度世界假设的白平衡算法实现色彩平衡的例子。

实例 3-22：使用基于灰度世界假设的白平衡算法实现色彩平衡
源码路径： daima\3\bai.py

```python
import cv2
import numpy as np

def white_balance(image):
    # 将图像转换为浮点数表示
    image = image.astype(float)

    # 计算每个颜色通道的平均值
    avg_R = np.mean(image[:, :, 2])
    avg_G = np.mean(image[:, :, 1])
    avg_B = np.mean(image[:, :, 0])

    # 计算平均灰度值
    avg_gray = (avg_R + avg_G + avg_B) / 3.0

    # 计算每个颜色通道的增益
    gain_R = avg_gray / avg_R
    gain_G = avg_gray / avg_G
    gain_B = avg_gray / avg_B

    # 对每个像素点进行颜色增益校正
    corrected_image = np.copy(image)
    corrected_image[:, :, 2] *= gain_R
    corrected_image[:, :, 1] *= gain_G
    corrected_image[:, :, 0] *= gain_B

    # 将像素值限制在 0～255 之间
    corrected_image = np.clip(corrected_image, 0, 255)

    # 将图像转换为 uint8 类型
    corrected_image = corrected_image.astype(np.uint8)

    return corrected_image

# 读取图像
image = cv2.imread('image.jpg', cv2.IMREAD_COLOR)
```

```
# 进行白平衡校正
balanced_image = white_balance(image)

# 显示原始图像和校正后的图像
cv2.imshow('Original Image', image)
cv2.imshow('Balanced Image', balanced_image)
cv2.waitKey(0)
cv2.destroyAllWindows()
```

在上述代码中，首先定义了一个 white_balance()函数来实现基于灰度世界假设的白平衡算法。该函数将图像转换为浮点数表示，并计算每个颜色通道的平均值。然后，根据平均灰度值和每个颜色通道的平均值之间的比例关系，计算出颜色增益。最后，根据颜色增益对每个像素点进行校正，得到色彩平衡的图像。代码执行后的效果如图 3-11 所示。

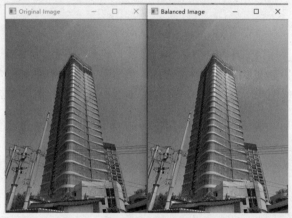

图 3-11 使用基于灰度世界假设的白平衡算法实现的效果

3.4.2 颜色校正

颜色校正（color correction）是一种通过调整图像的颜色分布来实现色彩平衡的方法。它可以用于校正图像中的色偏、色彩失真或颜色不一致的问题。颜色校正的目标是使图像的颜色看起来更加真实、自然和一致。

颜色校正的方法有很多种，其中一种常用的方法是直方图匹配（histogram matching）。直方图匹配通过将图像的颜色分布映射到一个目标颜色分布来实现颜色校正。这可以通过将原始图像的颜色直方图与目标颜色直方图进行比较和匹配来实现。例如，下面是一个使用直方图匹配实现颜色校正的例子。

实例 3-23：使用直方图匹配实现颜色校正

源码路径：daima\3\yan.py

```
import cv2
import numpy as np

def color_correction(image, target_hist):
    # 将图像转换为 Lab 颜色空间
    lab_image = cv2.cvtColor(image, cv2.COLOR_BGR2LAB)
    
    # 计算原始图像的颜色直方图
    original_hist, _ = np.histogram(lab_image[:, :, 1], bins=256, range=(0, 256))
    
    # 计算原始图像和目标直方图的累积分布函数
    original_cdf = original_hist.cumsum()
    original_cdf_normalized = original_cdf / original_cdf[-1]
    target_cdf_normalized = target_hist.cumsum() / target_hist.sum()
```

```python
    # 使用直方图匹配进行颜色校正
    lut = np.interp(original_cdf_normalized, target_cdf_normalized, np.arange(256))
    lab_image[:, :, 1] = np.interp(lab_image[:, :, 1], np.arange(256), lut)

    # 将图像转换回 BGR 颜色空间
    corrected_image = cv2.cvtColor(lab_image, cv2.COLOR_LAB2BGR)

    return corrected_image

# 读取原始图像和目标图像
original_image = cv2.imread('original_image.jpg', cv2.IMREAD_COLOR)
target_image = cv2.imread('target_image.jpg', cv2.IMREAD_COLOR)

# 将目标图像转换为 Lab 颜色空间并计算其颜色直方图
target_lab = cv2.cvtColor(target_image, cv2.COLOR_BGR2LAB)
target_hist, _ = np.histogram(target_lab[:, :, 1], bins=256, range=(0, 256))

# 进行颜色校正
corrected_image = color_correction(original_image, target_hist)

# 显示原始图像、目标图像和校正后的图像
cv2.imshow('Original Image', original_image)
cv2.imshow('Target Image', target_image)
cv2.imshow('Corrected Image', corrected_image)
cv2.waitKey(0)
cv2.destroyAllWindows()
```

在上述代码中，首先定义了一个 color_correction()函数来实现直方图匹配的颜色校正算法。该函数将图像转换为 Lab 颜色空间，计算原始图像的颜色直方图，并计算原始图像和目标图像直方图的累积分布函数。然后，使用 np.interp()函数将原始图像的颜色通道值映射到目标图像的颜色分布上，从而实现颜色校正。最后，将图像转换回 BGR 颜色空间，并得到校正后的图像。

通过直方图匹配进行颜色校正，可以使图像的颜色分布与目标图像的颜色分布相匹配，从而实现色彩平衡的效果。这个例子展示了一种复杂一点的颜色校正方法，通过调整图像的颜色分布，可以使图像看起来更加真实、自然和一致。

3.4.3 调整色调和饱和度

调整色调和饱和度（hue and saturation adjustment）通过调整图像的色调和饱和度参数来改变图像的颜色外观。可以通过调整色调曲线、饱和度增强等方法实现。校正图像的颜色偏移，以消除图像中的色温问题。常见的白平衡算法包括灰度世界假设、白点算法和基于颜色温度的算法。例如，下面是一个使用调整色调和饱和度实现色彩平衡的例子。

实例 3-24：使用调整色调和饱和度实现色彩平衡

源码路径： daima\3\tiao.py

```python
import cv2
import numpy as np

def color_balance(image, hue_shift, saturation_factor):
```

```python
# 将图像转换为 HSV 颜色空间
hsv_image = cv2.cvtColor(image, cv2.COLOR_BGR2HSV)

# 调整色调
hsv_image[:, :, 0] = (hsv_image[:, :, 0] + hue_shift) % 180

# 调整饱和度
hsv_image[:, :, 1] = np.clip(hsv_image[:, :, 1] * saturation_factor, 0, 255)

# 将图像转换回 BGR 颜色空间
balanced_image = cv2.cvtColor(hsv_image, cv2.COLOR_HSV2BGR)

return balanced_image

# 读取图像
image = cv2.imread('image.jpg', cv2.IMREAD_COLOR)

# 进行色彩平衡调整
hue_shift = 20  # 色调偏移量
saturation_factor = 1.5  # 饱和度因子
balanced_image = color_balance(image, hue_shift, saturation_factor)

# 显示原始图像和平衡后的图像
cv2.imshow('Original Image', image)
cv2.imshow('Balanced Image', balanced_image)
cv2.waitKey(0)
cv2.destroyAllWindows()
```

在上述代码中，首先定义了一个 color_balance()函数来实现调整色调和饱和度的色彩平衡方法。该函数首先将图像转换为 HSV 颜色空间，然后根据给定的色调偏移量和饱和度因子调整图像的色调和饱和度参数。最后，将图像转换回 BGR 颜色空间，并得到色彩平衡后的图像。

通过调整色调和饱和度，可以改变图像的整体色相和鲜艳程度，从而实现色彩平衡的效果。这个例子展示了一种稍微复杂一点的色彩平衡方法，通过调整色调和饱和度参数，可以使图像的颜色看起来更加平衡、鲜艳和自然。

3.5 超分辨率

超分辨率是一种图像处理技术，旨在通过增加图像的空间分辨率来实现图像增强。它可以从低分辨率图像中恢复出高分辨率的细节，从而提高图像的清晰度和细节可见性。超分辨率的实现方法有很多种，其中一种常用的方法是基于插值和图像重建的技术。下面是实现一个简要超分辨率的步骤：

（1）图像插值：使用插值算法（如双线性插值、双三次插值等）将低分辨率图像进行放大，以获得初始的高分辨率图像估计。

（2）图像重建：基于插值后的初始估计，使用图像重建算法（如基于边缘的重建、基于学习的重建等）来增加图像的细节和清晰度。这些算法通常通过利用图像的纹理特征和统计模型来进行重建。

（3）细节增强：对重建后的图像进行细节增强处理，以增强图像的细节和锐度。常用的方法包括锐化滤波、边缘增强等。

（4）后处理：对增强后的图像进行一些后处理步骤，例如去噪处理、色彩校正等，以进一步提升图像质量和视觉效果。

超分辨率技术在图像增强、图像重建、视频增强等领域都有广泛的应用。例如，下面是一个使用基于深度学习的超分辨率模型实现图像增强的例子，将使用 SRGAN（super-resolution generative adversarial network）模型来实现超分辨率。

实例 3-25：使用 SRGAN 模型来实现超分辨率
源码路径：daima\3\fenbian.py

```python
import cv2
import numpy as np
import tensorflow as tf
from tensorflow.keras.models import load_model

# 加载预训练的 SRGAN 模型
srgan_model = load_model('srgan_model.h5')

# 读取低分辨率图像
image = cv2.imread('low_resolution_image.jpg', cv2.IMREAD_COLOR)

# 将图像归一化到范围[0, 1]
image = image / 255.0

# 将图像转换为 Tensor 形式
image = tf.expand_dims(image, axis=0)

# 使用 SRGAN 模型进行超分辨率重建
reconstructed_image = srgan_model.predict(image)

# 将重建后的图像转换为 numpy 数组形式
reconstructed_image = np.squeeze(reconstructed_image) * 255.0

# 转换图像类型为 uint8
reconstructed_image = reconstructed_image.astype(np.uint8)

# 显示低分辨率图像和重建后的图像
cv2.imshow('Low Resolution Image', image)
cv2.imshow('Enhanced Image', reconstructed_image)
cv2.waitKey(0)
cv2.destroyAllWindows()
```

在上述代码中，首先加载了预训练的 SRGAN 模型（可通过训练过程得到）。然后，读取低分辨率的图像，并将其归一化到范围[0, 1]。接下来，使用 SRGAN 模型对低分辨率图像进行超分辨率重建，得到增强后的图像。最后，将增强后的图像转换为 uint8 类型，并显示低分辨率图像和重建后的图像。

本实例展示了一种使用基于深度学习的超分辨率模型实现图像增强的方法。通过使用训练有素的模型，可以从低分辨率图像中恢复出更多的细节，提高图像的清晰度和质量。此例子仅为概念演示，实际使用时需要适应自己的数据集和模型训练过程。

注意：超分辨率的结果受到原始低分辨率图像的限制，因此超分辨率并不能从低分辨率图像中恢复出所有丢失的细节，但可以在一定程度上提高图像的清晰度和细节可见性。

3.6 去除运动模糊

运动模糊是由于相机或拍摄对象的运动而导致的图像模糊效果。为了实现图像增强，可以采用去除运动模糊的方法，恢复图像的清晰度和细节。

3.6.1 边缘

基于边缘的方法（edge-based methods）是图像中的重要特征，可以用于恢复清晰图像。基于边缘的方法利用边缘检测和边缘保持算法来恢复边缘，从而提高图像的清晰度和细节可见性。使用基于边缘的方法的基本思想是通过利用图像中的边缘信息，对模糊图像进行分析和处理，以恢复原始图像的细节和清晰度。以下是使用基于边缘的方法去除运动模糊的简要过程：

（1）边缘检测：首先，使用边缘检测算法（如 sobel、canny 等）对模糊图像进行边缘检测，提取图像中的边缘信息。

（2）边缘增强：通过增强提取的边缘信息，突出边缘的细节和清晰度。这可以通过增加边缘的对比度、锐化边缘等方法来实现。

（3）逆运算：根据增强后的边缘信息，对模糊图像进行逆滤波或反卷积操作，以恢复原始图像的细节和清晰度。

注意，基于边缘的方法的具体实现和算法可能因应用场景和要求而有所差异，因此需要根据具体情况进行调整和改进。例如，下面是一个使用基于边缘的方法去除运动模糊的简单例子。

实例 3-26：使用基于边缘的方法去除运动模糊

源码路径：daima\3\bian.py

```python
import cv2
import numpy as np

def motion_deblur(image, kernel_size, motion_angle):
    # 生成运动模糊核
    kernel = np.zeros((kernel_size, kernel_size))
    center = kernel_size // 2
    kernel[center, :] = 1.0 / kernel_size
    # 对模糊核进行旋转
    M = cv2.getRotationMatrix2D((center, center), -motion_angle, 1.0)
    kernel = cv2.warpAffine(kernel, M, (kernel_size, kernel_size))

    # 进行逆滤波
    restored_image = cv2.filter2D(image, -1, np.linalg.pinv(kernel))

    return restored_image

# 读取模糊图像
image = cv2.imread('blurred_image.jpg', cv2.IMREAD_COLOR)

# 转换为灰度图像
gray_image = cv2.cvtColor(image, cv2.COLOR_BGR2GRAY)

# 进行边缘检测
edges = cv2.Canny(gray_image, 100, 200)
```

```
# 增强边缘信息
enhanced_edges = cv2.GaussianBlur(edges, (5, 5), 0)

# 进行逆运算恢复
restored_image = motion_deblur(enhanced_edges, 15, 45)

# 显示模糊图像、边缘图像和恢复后的图像
cv2.imshow('Blurred Image', gray_image)
cv2.imshow('Enhanced Edges', enhanced_edges)
cv2.imshow('Restored Image', restored_image)
cv2.waitKey(0)
cv2.destroyAllWindows()
```

上述代码演示了使用基于边缘的方法去除运动模糊的基本步骤。在实际应用中，可能需要根据具体情况对边缘信息进行更复杂的处理，以获得更好的结果。代码执行后的效果如图 3-12 所示。

图 3-12　基于边缘的方法去除运动模糊的效果

3.6.2　逆滤波

逆滤波（inverse filtering）是一种基本的去模糊方法，它通过计算模糊图像与逆滤波核的卷积来恢复清晰图像。逆滤波的效果受到噪声和伪影的影响，因此在实际应用中可能需要结合其他方法来改善结果。然而，逆滤波方法在实际应用中可能会面临一些挑战，例如噪声的增加和图像估计的不稳定性。因此，通常需要结合其他方法，如正则化技术或约束优化方法，来提高逆滤波的效果。下面将提供一个使用逆滤波方法去除运动模糊的例子。这个例子可能不是特别复杂，但可以展示逆滤波方法的基本原理。

实例 3-27：通过计算模糊图像与逆滤波核的卷积来恢复清晰图像
源码路径：daima\3\ni.py

```
import cv2
import numpy as np
```

```python
def motion_deblur(image, kernel_size, motion_angle):
    # 生成运动模糊核
    kernel = np.zeros((kernel_size, kernel_size))
    center = kernel_size // 2
    kernel[center, :] = 1.0 / kernel_size
    # 对模糊核进行旋转
    M = cv2.getRotationMatrix2D((center, center), -motion_angle, 1.0)
    kernel = cv2.warpAffine(kernel, M, (kernel_size, kernel_size))

    # 进行逆滤波
    restored_image = cv2.filter2D(image, -1, np.linalg.pinv(kernel))

    return restored_image

# 读取模糊图像
image = cv2.imread('blurred_image.jpg', cv2.IMREAD_COLOR)

# 转换为灰度图像
gray_image = cv2.cvtColor(image, cv2.COLOR_BGR2GRAY)

# 进行逆滤波恢复
kernel_size = 15     # 模糊核大小
motion_angle = 45    # 运动方向（逆时针旋转角度）
restored_image = motion_deblur(gray_image, kernel_size, motion_angle)

# 显示模糊图像和恢复后的图像
cv2.imshow('Blurred Image', gray_image)
cv2.imshow('Restored Image', restored_image)
cv2.waitKey(0)
cv2.destroyAllWindows()
```

在上述代码中，首先读取模糊图像，并将其转换为灰度图像。然后，通过指定模糊核的大小和运动方向，使用逆滤波方法对图像进行恢复。最后，显示模糊图像和恢复后的图像。根据需要调整 kernel_size 和 motion_angle 的值。注意，逆滤波方法在处理真实世界的复杂模糊情况时可能效果不理想，因此可能需要结合其他技术或算法来进一步改进结果。

3.6.3 统计方法

统计方法（statistical methods）是利用多个模糊图像或先验知识进行建模和估计，以恢复清晰图像。这些方法基于图像的统计特性和概率模型，例如最大似然估计、最小二乘法等。方法的基本思想是通过对模糊图像中的像素值进行统计分析，推断出运动模糊的参数，并进行逆运算来恢复原始图像。以下是使用统计方法去除运动模糊的简要过程：

（1）统计分析：对模糊图像中的像素值进行统计分析，例如利用图像中的边缘信息或图像梯度信息来推断运动模糊的方向和程度。

（2）参数估计：基于统计分析的结果，估计运动模糊的参数，如模糊核的长度和方向。

（3）逆运算：根据估计的运动模糊参数，对模糊图像进行逆滤波或反卷积操作，尽可能还原原始图像的细节。

注意，统计方法的具体实现和算法可能因应用场景和要求而有所差异，因此需要根据具体情况进行调整和改进。例如，下面是一个使用统计方法去除运动模糊的简单例子。

实例 3-28：使用统计方法去除运动模糊

源码路径：daima\3\jin.py

```python
import cv2
import numpy as np

def motion_deblur(image, kernel_size, motion_angle):
    # 生成运动模糊核
    kernel = np.zeros((kernel_size, kernel_size))
    center = kernel_size // 2
    kernel[center, :] = 1.0 / kernel_size
    # 对模糊核进行旋转
    M = cv2.getRotationMatrix2D((center, center), -motion_angle, 1.0)
    kernel = cv2.warpAffine(kernel, M, (kernel_size, kernel_size))

    # 进行逆滤波
    restored_image = cv2.filter2D(image, -1, np.linalg.pinv(kernel))

    return restored_image

# 读取模糊图像
image = cv2.imread('blurred_image.jpg', cv2.IMREAD_COLOR)

# 转换为灰度图像
gray_image = cv2.cvtColor(image, cv2.COLOR_BGR2GRAY)

# 进行统计分析和参数估计
kernel_size = 15  # 模糊核大小
motion_angle = 45  # 运动方向（逆时针旋转角度）

# 进行逆运算恢复
restored_image = motion_deblur(gray_image, kernel_size, motion_angle)

# 显示模糊图像和恢复后的图像
cv2.imshow('Blurred Image', gray_image)
cv2.imshow('Restored Image', restored_image)
cv2.waitKey(0)
cv2.destroyAllWindows()
```

在上述代码中，需要读者根据需要调整 kernel_size 和 motion_angle 的值。这个例子是一个简化的示例，演示了使用统计方法去除运动模糊的基本原理。

3.6.4 盲去卷积

基于盲去卷积（blind deconvolution）的方法是一种无须事先知道模糊核的方法。盲去卷积方法需要较高的计算复杂度，并且对于复杂的模糊情况可能存在困难。盲去卷积方法的核心思想是通过迭代优化过程来估计模糊核和清晰图像，以最小化重建图像与模糊图像之间的差异。例如，下面是一个使用盲去卷积方法去除运动模糊的例子。

实例 3-29：使用盲去卷积方法去除运动模糊

源码路径：daima\3\mang.py

```python
import cv2
import numpy as np
```

```python
from scipy.signal import convolve2d

def blind_deconvolution(image, kernel_size, iterations):
    # 初始化模糊核和清晰图像
    kernel = np.zeros((kernel_size, kernel_size))
    kernel[kernel_size//2, :] = 1.0 / kernel_size

    # 盲去卷积迭代过程
    for _ in range(iterations):
        # 估计模糊图像
        blurred_image = convolve2d(image, kernel, mode='same', boundary='symm', fillvalue=0)

        # 更新模糊核
        restored_image = convolve2d(blurred_image, np.rot90(kernel, 2), mode='same', boundary='symm', fillvalue=0)

    return restored_image

# 读取模糊图像
image = cv2.imread('blurred_image.jpg', cv2.IMREAD_COLOR)

# 转换为灰度图像
gray_image = cv2.cvtColor(image, cv2.COLOR_BGR2GRAY)

# 进行盲去卷积恢复
kernel_size = 15  # 模糊核大小
iterations = 10  # 迭代次数
restored_image = blind_deconvolution(gray_image, kernel_size, iterations)

# 显示模糊图像和恢复后的图像
cv2.imshow('Blurred Image', gray_image)
cv2.imshow('Restored Image', restored_image)
cv2.waitKey(0)
cv2.destroyAllWindows()
```

使用运动模糊的盲去卷积方法仅适用于特定类型的模糊,并且可能需要根据实际情况进行调整以获得更好的结果。

第 4 章　图像特征提取处理

图像特征提取是计算机视觉和图像处理领域的重要任务，它是指从图像数据中提取有意义的、可用于表征和描述图像内容的信息。这些特征可以用于图像分类、目标检测、图像匹配、图像检索等应用中。本章详细讲解使用 Python 语言实现图像特征提取的知识。

4.1　图像特征提取方法

在现实应用中，有如下几种常用的图像特征提取方法：
- 颜色特征：颜色是图像中重要的信息之一。常见的颜色特征提取方法包括直方图颜色特征和颜色矩。直方图颜色特征统计图像中各个颜色通道的像素数量分布，用于表示图像的整体颜色分布情况。颜色矩是对颜色分布的统计特征，包括均值、方差、协方差等。
- 纹理特征：纹理特征描述图像中的纹理结构，用于表征图像的细节信息。常用的纹理特征提取方法包括灰度共生矩阵（GLCM）、局部二值模式（LBP）、gabor 滤波器等。GLCM 统计图像中不同灰度级别的像素对出现的概率，用于描述图像的纹理统计特性。LBP 对每个像素点计算局部二值模式，并统计不同模式的出现频率，用于表示图像的纹理信息。gabor 滤波器是一组带有不同频率和方向的滤波器，用于提取图像的纹理特征。
- 形状特征：形状特征描述图像中对象的形状和轮廓信息。常用的形状特征包括轮廓特征、边界框特征和几何矩。轮廓特征描述对象的边界形状，可以使用轮廓的长度、面积、周长等进行表征。边界框特征是利用对象的最小外接矩形或最小外接圆来描述对象的形状。几何矩是对图像像素位置的统计量，用于表示图像的形状和几何特性。
- 尺度不变特征变换（SIFT）：SIFT 是一种具有尺度和旋转不变性的特征提取算法。它通过在不同尺度和方向上检测和描述局部特征，生成具有唯一性和稳定性的特征描述子。SIFT 特征对图像的缩放、旋转、平移等变换具有较好的鲁棒性，广泛应用于图像匹配和目标检测领域。

上述方法只是图像特征提取的一部分，另外还有很多其他的特征提取方法，如边缘特征、角点特征、HOG 特征等。根据具体应用和需求，可以选择适合的特征提取方法，或者结合多个方法进行综合特征表示。特征提取是计算机视觉和图像处理中的基础工作，合理选择和设计特征可以对后续的图像分析和处理任务产生重要影响。

4.2　颜色特征

使用颜色特征方法进行图像特征提取是一种常见的计算机视觉技术，用于描述图像中的颜色信息。通过提取图像的颜色特征，可以用于图像分类、检索、目标识别等应用。

4.2.1 颜色直方图

颜色直方图是一种统计图像中各种颜色出现频率的方法。它将图像的颜色空间分成若干个颜色通道（如 RGB 通道或 HSV 通道），并统计每个通道中每个颜色的像素数量。颜色直方图可以用于描述图像的颜色分布情况，反映了图像中不同颜色的数量和分布比例。通过计算颜色直方图，可以得到一个向量表示图像的颜色特征。

颜色直方图的计算过程包括将图像转换到指定的颜色空间，将颜色空间划分为若干个区间，然后统计每个区间内的像素数量。在 Python 程序中，可以使用 NumPy 和 OpenCV 等库实现颜色直方图，例如下面的实例。

实例 4-1：使用 NumPy 和 OpenCV 等库实现颜色直方图
源码路径：daima\4\yanzhi.py

```python
import cv2
import numpy as np
from sklearn.cluster import KMeans

def extract_main_colors(image, num_colors):
    # 将图像转换为 RGB 颜色空间
    image = cv2.cvtColor(image, cv2.COLOR_BGR2RGB)

    # 将图像从三维数组转换为二维数组
    pixels = image.reshape(-1, 3)

    # 使用 K 均值聚类算法提取主要颜色
    kmeans = KMeans(n_clusters=num_colors)
    kmeans.fit(pixels)

    # 获取聚类中心（主要颜色）
    main_colors = kmeans.cluster_centers_

    return main_colors.astype(np.uint8)

# 读取图像
image = cv2.imread('image.jpg')

# 提取图像中的主要颜色
num_colors = 5
main_colors = extract_main_colors(image, num_colors)

# 显示主要颜色
for color in main_colors:
    color = np.array([[color]], dtype=np.uint8)
    color_image = cv2.cvtColor(color, cv2.COLOR_RGB2BGR)
    cv2.imshow("Main Color", color_image)
    cv2.waitKey(0)

cv2.destroyAllWindows()
```

在上述代码中，首先将图像从 BGR 颜色空间转换为 RGB 颜色空间。然后，将图像的像素值重新排列为一个二维数组，每行表示一个像素点的 RGB 值。接下来，使用 K 均值聚类算法对像素进行聚类，指定要提取的主要颜色数量。最后，获取聚类中心作为主要颜色，并将其显示出来。

执行上述代码后，程序将显示提取出的图像的主要颜色。每个主要颜色都会以独立的窗口显示出来。这时，可以按下任意键来逐个查看主要颜色窗口，按 Esc 键退出程序。

4.2.2 其他颜色特征提取方法

除了颜色直方图外，还有其他一些颜色特征提取方法，如颜色矩、颜色共生矩阵等。这些方法可以进一步细化对图像颜色分布的描述，从而获得更丰富的颜色特征。

1．颜色矩

当提取图像特征时，颜色矩是一种常用的方法。颜色矩是一种用于描述图像颜色分布的统计特征。它可以提供关于图像颜色分布的信息，例如平均颜色、颜色的分散程度等。通过计算颜色矩，可以获得对图像进行分类、检索和识别等任务非常有用的特征。例如，下面是一个使用颜色矩方法实现图像特征提取的例子，其中包括计算颜色矩的过程。

实例 4-2：使用颜色矩方法实现图像特征提取

源码路径：daima\4\yansegui.py

```python
import cv2
import numpy as np

def calculate_color_moments(image):
    # 将图像转换为 HSV 颜色空间
    hsv_image = cv2.cvtColor(image, cv2.COLOR_BGR2HSV)

    # 分割 HSV 图像的通道
    h, s, v = cv2.split(hsv_image)

    # 计算颜色矩
    h_mean = np.mean(h)
    s_mean = np.mean(s)
    v_mean = np.mean(v)
    h_std = np.std(h)
    s_std = np.std(s)
    v_std = np.std(v)

    return h_mean, s_mean, v_mean, h_std, s_std, v_std

# 读取图像
image = cv2.imread('image.jpg')

# 计算图像的颜色矩
h_mean, s_mean, v_mean, h_std, s_std, v_std = calculate_color_moments(image)

# 打印颜色矩的值
print("Hue Mean:", h_mean)
print("Saturation Mean:", s_mean)
print("Value Mean:", v_mean)
print("Hue Standard Deviation:", h_std)
print("Saturation Standard Deviation:", s_std)
print("Value Standard Deviation:", v_std)
```

在上述代码中，首先将图像从 BGR 颜色空间转换为 HSV 颜色空间。然后，将 HSV 图像的通道分离为独立的图像数组。接下来，计算每个通道的颜色矩，包括均值和标准差。最后，打印出计算得到的颜色矩值。执行后会输出：
```
Hue Mean: 104.417625
Saturation Mean: 92.099975
Value Mean: 184.59414166666667
Hue Standard Deviation: 19.779323742047108
Saturation Standard Deviation: 37.293618488950294
Value Standard Deviation: 52.250336400957885
```

颜色矩计算的结果将提供关于图像颜色分布的统计信息。在这个例子中，计算了hue（色调）、saturation（饱和度）和value（亮度）通道的均值和标准差。这些值可以用于描述图像的整体颜色特征。

注意：这个例子演示了如何使用颜色矩方法提取图像的颜色特征。可以根据需要扩展代码，计算更多颜色通道的矩特征，或者将颜色矩与其他特征描述符结合使用，以实现更复杂的图像分析任务。

2．颜色共生矩阵

颜色共生矩阵（color co-occurrence matrix，CCM）是一种常用的图像特征提取方法。颜色共生矩阵描述了图像中不同颜色对的出现频率和位置关系，可以提供关于纹理和颜色分布的信息。通过计算颜色共生矩阵，我们可以获得用于分类、检索和识别等任务的有效特征。例如，下面是一个使用颜色共生矩阵方法实现图像特征提取的例子，其中包括计算颜色共生矩阵的过程。

实例4-3：使用颜色共生矩阵方法实现图像特征提取

源码路径：daima\4\juzhen.py

```python
import cv2
import numpy as np
from skimage.feature import greycomatrix, greycoprops

def calculate_glcm(image, distance, angle):
    # 将图像转换为灰度图像
    gray_image = cv2.cvtColor(image, cv2.COLOR_BGR2GRAY)
    
    # 计算颜色共生矩阵
    glcm = greycomatrix(gray_image, [distance], [angle], levels=256, symmetric=True, normed=True)
    
    return glcm

# 读取图像
image = cv2.imread('image.jpg')

# 计算颜色共生矩阵
glcm = calculate_glcm(image, distance=1, angle=0)

# 计算颜色共生矩阵的某些特征
contrast = greycoprops(glcm, 'contrast')
dissimilarity = greycoprops(glcm, 'dissimilarity')
homogeneity = greycoprops(glcm, 'homogeneity')

# 打印计算得到的特征值
print('Contrast:', contrast)
print('Dissimilarity:', dissimilarity)
```

```
print('Homogeneity:', homogeneity)
```

在上述代码中，首先将图像从 BGR 颜色空间转换为灰度图像。然后，使用模块 skimage.feature 中的 greycomatrix 函数来计算颜色共生矩阵。注意，我们还使用 greycoprops 函数来计算颜色共生矩阵的一些特征，如对比度（contrast）、不相似度（dissimilarity）和均匀性（homogeneity）。执行会输出：

```
Contrast: [[427.90448161]]
Dissimilarity: [[9.28667224]]
Homogeneity: [[0.47605069]]
```

注意：本节介绍的颜色特征方法的实现相对复杂，需要对图像进行一些预处理和特征提取计算。在实际应用中，可以根据具体任务选择适合的颜色特征提取方法，并结合其他特征进行综合描述和分析。

4.3 纹理特征

纹理特征作为图像特征提取的依据，它主要关注图像中的纹理和结构信息。纹理特征可以帮助我们捕捉到图像中的细节、重要的纹理模式和结构信息，从而用于图像分类、目标检测、图像匹配等任务。常用的纹理特征提取方法有灰度共生矩阵、方向梯度直方图、尺度不变特征变换和小波变换等。上述纹理特征提取方法可以通过使用相应的库和算法进行实现。在 Python 程序中，可以使用 OpenCV、scikit-image、PyWavelets 等库来实现这些方法。具体实现的步骤和参数设置会根据不同的方法而有所差异。开发者需要根据具体的应用场景和任务选择适合的纹理特征提取方法，并根据实际情况调整参数和处理步骤，以获得更好的图像特征表示。

4.3.1 灰度共生矩阵

灰度共生矩阵（gray-level co-occurrence matrix，GLCM）是一种常用的图像纹理特征提取技术，是一种描述图像中像素灰度级之间相对关系的矩阵，用于描述图像中像素对之间的灰度值共生关系，通过统计相邻像素的灰度值出现频次和空间关系，从而提取出图像的纹理特征。灰度共生矩阵通过计算相邻像素灰度值的统计特性，如共生矩阵的对比度、能量、熵等，可以提取出图像的纹理特征。

下面是使用灰度共生矩阵方法实现图像纹理特征提取的步骤：

（1）将彩色图像转换为灰度图像。由于灰度共生矩阵方法是基于灰度图像的，所以首先需要将彩色图像转换为灰度图像。

（2）定义灰度共生矩阵参数。包括灰度级数目、灰度共生矩阵的距离和方向等参数。灰度级数目表示将灰度值分为多少个等级；距离表示计算灰度共生矩阵时像素对之间的距离；方向表示计算灰度共生矩阵时像素对之间的方向。

（3）计算灰度共生矩阵。遍历图像的每个像素点，对于每个像素点，计算与其相邻像素点的灰度值关系，统计出现频次，并更新灰度共生矩阵。

例如，下面是一个使用灰度共生矩阵方法提取纹理特征的实例。

实例 4-4：使用灰度共生矩阵方法提取纹理特征

源码路径：daima\4\huigong.py

```python
import cv2
from skimage.feature import greycomatrix, greycoprops

def calculate_glcm(image, distances, angles):
    gray_image = cv2.cvtColor(image, cv2.COLOR_BGR2GRAY)
    glcm = greycomatrix(gray_image, distances, angles, levels=256, symmetric=True, normed=True)
    return glcm

def extract_texture_features(glcm):
    contrast = greycoprops(glcm, 'contrast')
    energy = greycoprops(glcm, 'energy')
    correlation = greycoprops(glcm, 'correlation')
    homogeneity = greycoprops(glcm, 'homogeneity')
    return contrast, energy, correlation, homogeneity

# 读取图像
image = cv2.imread('image.jpg')

# 计算灰度共生矩阵
distances = [1]  # 距离
angles = [0]  # 方向
glcm = calculate_glcm(image, distances, angles)

# 提取纹理特征
contrast, energy, correlation, homogeneity = extract_texture_features(glcm)

# 打印纹理特征
print('Contrast:', contrast)
print('Energy:', energy)
print('Correlation:', correlation)
print('Homogeneity:', homogeneity)
```

在上述代码中，首先读取图像，然后使用 calculate_glcm()函数计算图像的灰度共生矩阵。接下来，使用 extract_texture_features()函数提取各种纹理特征，如对比度（contrast）、能量（energy）、相关性（correlation）和均匀性（homogeneity）。最后，打印出这些纹理特征的值。执行后会输出：

```
Contrast: [[427.90448161]]
Energy: [[0.05509271]]
Correlation: [[0.85237939]]
Homogeneity: [[0.47605069]]
```

4.3.2 方向梯度直方图

方向梯度直方图（histogram of oriented gradients，HOG）是一种基于梯度信息的纹理特征提取方法，它通过计算图像中各个像素点的梯度方向和梯度强度，然后将图像划分为小的区域，统计每个区域内不同梯度方向的像素数量，最终形成一个直方图来表示图像的纹理特征。HOG 特征提取的基本步骤如下：

（1）将图像转换为灰度图像，以便计算梯度信息。

（2）对图像进行局部梯度计算，通常使用 sobel 算子或其他梯度算子。
（3）将图像划分为小的局部区域（cell），对每个区域内的梯度进行统计。
（4）将局部区域组合成更大的块（block），对每个块内的局部区域梯度进行归一化和组合。
（5）构建方向梯度直方图，将每个块的梯度信息组合成一个特征向量。

例如，下面是一个使用 Python 和 scikit-image 库实现 HOG 特征提取的例子。

实例 4-5：使用 Python 和 scikit-image 库实现 HOG 特征提取
源码路径：daima\4\hog.py

```
import cv2
from skimage.feature import hog

def extract_hog_features(image):
    gray_image = cv2.cvtColor(image, cv2.COLOR_BGR2GRAY)
    hog_features = hog(gray_image, orientations=9, pixels_per_cell=(8, 8), cells_per_block=(2, 2), block_norm='L2-Hys')
    return hog_features

# 读取图像
image = cv2.imread('image.jpg')

# 提取 HOG 特征
hog_features = extract_hog_features(image)

# 打印 HOG 特征向量
print('HOG Features:', hog_features)
```

在上述代码中，首先读取图像，然后使用 extract_hog_features()函数提取图像的 HOG 特征。在 HOG 函数中，我们指定了 9 个方向的梯度，每个 cell 的大小为 8×8 像素，每个 block 包含 2×2 个 cell，并使用 L2-Hys 归一化方式。最后，打印出 HOG 特征向量。执行后会输出：

```
HOG Features: [0.37306755 0.04656909 0.11781152 ... 0.02256336 0.00481969 0.00584473]
```

HOG 特征可以用于图像分类、目标检测和行人识别等场景，它对于描述图像的纹理和形状信息具有较好的性能。当然，具体的应用和参数设置可以根据任务的需求进行调整和优化。

4.3.3 尺度不变特征变换

尺度不变特征变换（scale-invariant feature transform，SIFT）是一种局部特征提取算法，它能够在图像中检测到关键点，并提取出与这些关键点相关的局部纹理特征。SIFT 算法通过在不同尺度和方向上对图像进行高斯滤波和梯度计算，然后使用局部图像块的特征描述子来表示图像的纹理特征。SIFT 特征提取的基本步骤如下：

（1）尺度空间极值点检测。在不同尺度空间中，通过 DoG 金字塔来寻找图像的极值点，即关键点（keypoint）。

（2）关键点定位。对检测到的关键点进行精确定位，排除低对比度和边缘响应较大的关键点。

（3）方向分配。为每个关键点分配主方向，用于后续的特征描述。

（4）特征描述。在每个关键点的周围区域内计算局部特征向量，该向量具有尺度和旋转不变性。

例如，下面是一个使用 Python 和 OpenCV 库实现 SIFT 特征提取的例子。

实例 4-6：使用 Python 和 OpenCV 库实现 SIFT 特征提取

源码路径： daima\4\chi.py

```
import cv2

def extract_sift_features(image):
    gray_image = cv2.cvtColor(image, cv2.COLOR_BGR2GRAY)
    sift = cv2.SIFT_create()
    keypoints, descriptors = sift.detectAndCompute(gray_image, None)
    return keypoints, descriptors

# 读取图像
image = cv2.imread('image.jpg')

# 提取 SIFT 特征
keypoints, descriptors = extract_sift_features(image)

# 在图像上绘制关键点
image_with_keypoints = cv2.drawKeypoints(image, keypoints, None)

# 显示图像和关键点
cv2.imshow("Image with Keypoints", image_with_keypoints)
cv2.waitKey(0)
cv2.destroyAllWindows()
```

在上述代码中，首先读取图像，然后使用 extract_sift_features()函数提取图像的 SIFT 特征。在函数中，我们将图像转换为灰度图像，然后创建 SIFT 对象。通过 detectAndCompute()函数可以同时检测关键点并计算对应的描述符。最后，使用 drawKeypoints()函数将关键点绘制在图像上，并显示结果。代码执行后的效果如图 4-1 所示。

SIFT 特征在图像匹配、目标跟踪和图像拼接中具有广泛的应用，它能够提取出具有尺度和旋转不变性的稳定特征点，对于处理具有视角变化和尺度变化的图像数据非常有用。

图 4-1 使用 Python 和 OpenCV 库实现 SIFT 特征提取的效果

4.3.4 小波变换

小波变换是一种用于分析信号和图像的数学工具，可以在不同频率和尺度上对信号进行分解和表示。小波变换可以用于图像纹理特征提取，通过分解图像的频域信息和空域信息，从而获取到不同尺度和方向上的纹理特征。小波变换是一种多尺度分析方法，它能够将图像分解为不同尺度和频率的子图像。在小波变换的过程中，可以提取出图像的纹理特征，例如局部纹理的频率、方向和能量等信息。

实现小波变换的基本步骤如下：

（1）选择合适的小波基函数。小波基函数是用来分析信号的基础函数，常用的有 haar 小波、

db 小波等。

（2）进行多尺度分解。将图像通过小波基函数进行多尺度分解，得到图像在不同频率和尺度上的分量。

（3）提取纹理特征。根据不同尺度和方向上的小波系数，可以提取出图像的纹理特征，如纹理的粗细、方向、对比度等。

（4）重构图像。根据提取的特征，可以进行逆小波变换，将图像重构回原始图像空间。

下面是一个使用 Python 和 PyWavelets 库实现小波变换图像纹理特征提取的例子。

实例 4-7：使用 PyWavelets 库实现小波变换图像纹理特征提取

源码路径：daima\4\bo.py

```python
import cv2
import pywt

def extract_texture_features(image):
    # 将图像转为灰度图
    gray_image = cv2.cvtColor(image, cv2.COLOR_BGR2GRAY)

    # 进行小波变换
    coeffs = pywt.dwt2(gray_image, 'haar')
    cA, (cH, cV, cD) = coeffs

    # 提取纹理特征
    texture_features = {
        'approximation': cA,
        'horizontal_detail': cH,
        'vertical_detail': cV,
        'diagonal_detail': cD
    }

    return texture_features

# 读取图像
image = cv2.imread('image.jpg')

# 提取纹理特征
texture_features = extract_texture_features(image)

# 显示原始图像和纹理特征
cv2.imshow("Original Image", image)
cv2.imshow("Approximation", texture_features['approximation'])
cv2.imshow("Horizontal Detail", texture_features['horizontal_detail'])
cv2.imshow("Vertical Detail", texture_features['vertical_detail'])
cv2.imshow("Diagonal Detail", texture_features['diagonal_detail'])
cv2.waitKey(0)
cv2.destroyAllWindows()
```

在上述代码中，首先将图像转换为灰度图像，然后使用 PyWavelets 库中的 dwt2()函数进行小波变换。选择 haar 小波作为基函数，通过分解得到近似系数（approximation）和细节系数（horizontal_detail、vertical_detail、diagonal_detail）。这些细节系数表示了图像在不同尺度和方向上的纹理信息。最后，使用 imshow()函数将原始图像和纹理特征展示出来，可以观察到不同细节系数所表达的纹理特征。代码执行后的效果如图 4-2 所示。

小波变换在图像纹理分析、纹理识别、图像压缩等领域有广泛应用,它能够提取图像的多尺度、多方向的纹理特征。

图 4-2 使用 PyWavelets 库实现小波变换图像纹理特征提取的效果

4.4 形状特征

形状特征是用于描述图像或物体形状的特征,它们可以用于图像分析、目标检测、图像识别和计算机视觉等领域。形状特征提取的目标是从图像中提取出能够描述物体形状的信息,以便对物体进行识别、分类或测量。常用的形状特征提取方法有:边界描述子、预处理后的轮廓特征、模型拟合方法、形状上的变换。

4.4.1 边界描述子

边界描述子是一种常用的形状特征提取方法,它通过对物体的边界进行分析和描述,从中提取出能够描述形状的特征。下面详细介绍边界描述子的原理,并给出两个实用且稍微复杂的例子。边界描述子的原理如下:

(1)获取物体的边界。首先需要获取物体的边界,可以通过边缘检测算法(如 canny 边缘检测)或轮廓检测算法(如 OpenCV 中的 findContours 函数)来获得物体的边界。

(2)归一化边界。对于获取的边界点集,将其进行归一化处理,使得边界的起点为坐标原点,同时进行平移和缩放操作,使得边界点分布在一个固定的区域内。

(3)提取边界描述子。对归一化后的边界点集进行特征提取。常见的边界描述子包括:傅里叶描述子和形状上下文。傅里叶描述子(fourier descriptors):将归一化的边界点集进行傅里叶变换,提取频域特征。傅里叶描述子可以用于对边界形状的旋转、缩放和平移具有不变性。形状上下文(shape context):通过计算边界点与其他点之间的相对位置关系,构建形状上下文描述子。形状上下文描述子可以用于对边界形状的旋转和尺度具有不变性。

(4)应用边界描述子。提取的边界描述子可以用于形状匹配、物体识别和分类等任务。

在现实应用中,可以通过如下两种方法实现边界描述子在形状特征提取中的应用。

1. 使用傅里叶描述子进行形状匹配

假设有一组图像中的物体边界,我们想要在新的图像中识别相似形状的物体,基本步骤如下:

(1)对于每个图像的物体边界,应用边缘检测算法获取边界点集。
(2)对边界点集进行归一化处理。
(3)对归一化后的边界点集计算傅里叶描述子。
(4)在新的图像中,提取物体边界并进行归一化处理。
(5)对新图像的归一化边界点集计算傅里叶描述子。
(6)对比新图像的傅里叶描述子与之前图像的傅里叶描述子,使用相似度度量方法(如欧氏距离)进行形状匹配。

例如,下面是一个使用 Python 语言实现傅里叶描述子形状匹配的简单例子。

实例 4-8:实现傅里叶描述子形状匹配
源码路径:daima\4\foliye.py

```python
import cv2
import numpy as np

def calculate_fourier_descriptor(image):
    # 提取轮廓
    contours, _ = cv2.findContours(image, cv2.RETR_EXTERNAL, cv2.CHAIN_APPROX_NONE)
    contour = contours[0]  # 假设只有一个轮廓

    # 计算傅里叶描述子
    contour_complex = np.empty(contour.shape[:-1], dtype=complex)
    contour_complex.real = contour[:, 0, 0]
    contour_complex.imag = contour[:, 0, 1]
    fourier_descriptor = np.fft.fft(contour_complex)

    return fourier_descriptor

# 读取数据库图像和查询图像
database_image = cv2.imread('database.jpg', cv2.IMREAD_GRAYSCALE)
query_image = cv2.imread('query.jpg', cv2.IMREAD_GRAYSCALE)

# 预处理图像(二值化等)
_, database_image = cv2.threshold(database_image, 127, 255, cv2.THRESH_BINARY)
_, query_image = cv2.threshold(query_image, 127, 255, cv2.THRESH_BINARY)

# 计算数据库图像和查询图像的傅里叶描述子
database_descriptor = calculate_fourier_descriptor(database_image)
query_descriptor = calculate_fourier_descriptor(query_image)

# 计算傅里叶描述子之间的距离
distance = np.linalg.norm(database_descriptor - query_descriptor)

print("Distance:", distance)
```

在上述代码中,请确保已准备好两幅图像作为数据库图像和查询图像,并将其命名为 database.jpg 和 query.jpg。此代码将计算数据库图像和查询图像的傅里叶描述子,并计算描述子之间的欧氏距离作为形状匹配的度量。注意,此示例仍然假设图像中只有一个轮廓。

2. 使用形状上下文描述子进行手势识别
假设有一组手势的图像,我们想要对新的手势图像进行识别,基本步骤如下:
(1)对于每个手势图像,应用边缘检测算法获取边界点集。

（2）对边界点集进行归一化处理。
（3）对归一化后的边界点集计算形状上下文描述子。
（4）在新的手势图像中，提取边界并进行归一化处理。
（5）对新图像的归一化边界点集计算形状上下文描述子。
（6）对比新图像的形状上下文描述子与之前手势图像的描述子，使用相似度度量方法（如相关系数）进行手势识别。

例如，下面是一个使用 HOG 特征和支持向量机（support vector machine，SVM）实现手势识别的例子。

实例 4-9：使用 HOG 特征和支持向量机实现手势识别

源码路径：daima\4\xing.py

```python
import cv2
import numpy as np
from sklearn.svm import SVC
from skimage.feature import hog
from skimage import data, exposure

# 加载手势图像数据集
gesture_images = []
gesture_labels = []

for i in range(1, 6):
    image = cv2.imread(f'gesture_{i}.jpg', cv2.IMREAD_GRAYSCALE)
    gesture_images.append(image)
    gesture_labels.append(i)

# 提取手势图像的 HOG 特征
gesture_hogs = []

for image in gesture_images:
    # 计算 HOG 特征
    hog_features, hog_image = hog(image, orientations=9, pixels_per_cell=(8, 8),
                            cells_per_block=(2, 2), visualize=True)
    # 对 HOG 图像进行直方图均衡化，增强可视化效果
    hog_image = exposure.rescale_intensity(hog_image, in_range=(0, 10))
    gesture_hogs.append(hog_features)

# 创建 SVM 分类器
svm = SVC()

# 使用 HOG 特征训练分类器
svm.fit(gesture_hogs, gesture_labels)

# 加载待识别手势图像
test_image = cv2.imread('test_gesture.jpg', cv2.IMREAD_GRAYSCALE)

# 提取待识别手势图像的 HOG 特征
test_hog = hog(test_image, orientations=9, pixels_per_cell=(8, 8),
            cells_per_block=(2, 2))

# 使用 SVM 分类器进行手势识别
predicted_label = svm.predict([test_hog])
```

```
print("Predicted Label:", predicted_label[0])
```

在上述代码中使用库 skimage 的 hog()函数来提取手势图像的 HOG 特征。然后，使用这些特征和对应的标签训练了一个 SVM 分类器。接下来，加载待识别的手势图像，提取其 HOG 特征，并使用 SVM 分类器进行手势识别，得到预测标签。请确保准备了手势图像数据集，并将手势图像命名为 gesture_1.jpg、gesture_2.jpg 等，将待识别的手势图像命名为 test_gesture.jpg。此代码将根据 HOG 特征进行手势识别，并输出预测的手势标签。

通过边界描述子的提取和匹配，可以实现对具有相似形状的物体或手势进行识别和分类。这些例子展示了边界描述子在形状特征提取中的应用，具有实用性、有趣性和一定的复杂性。

4.4.2　预处理后的轮廓特征

预处理后的轮廓特征是一种常用的图像特征提取方法，它通过对图像进行预处理和轮廓提取，然后分析轮廓的形状、大小、方向等特征来描述图像的形状和结构。预处理后的轮廓特征基于对物体边界进行预处理，以减少噪声和不相关信息。常用的方法有：

1. 链码

链码（chain code）是一种用于形状描述和特征提取的方法，它将轮廓视为一系列相邻的像素点的有序序列，通过记录像素点之间的连接顺序来表示轮廓的形状。链码方法具有简洁、紧凑的表示形式，适用于描述闭合轮廓的形状特征。将边界转换为链码，描述连续的边界点之间的连接关系。例如，下面是一个使用轮廓近似方法（cv2.approxPolyDP）实现链码特征提取的例子。

实例 4-10：使用轮廓近似方法（cv2.approxPolyDP）实现链码特征提取
源码路径：daima\4\lian.py

```
import cv2
import math

# 读取图像并转为灰度图像
image = cv2.imread('884.jpg')
gray = cv2.cvtColor(image, cv2.COLOR_BGR2GRAY)

# 二值化处理
_, binary = cv2.threshold(gray, 127, 255, cv2.THRESH_BINARY)

# 查找轮廓
contours, _ = cv2.findContours(binary, cv2.RETR_EXTERNAL, cv2.CHAIN_APPROX_NONE)

# 获取最长轮廓
longest_contour = max(contours, key=len)

# 使用链码获取形状特征
epsilon = 0.02 * cv2.arcLength(longest_contour, True)
chain_code = cv2.approxPolyDP(longest_contour, epsilon, True)


# 计算距离直方图
max_distance = 0
distance_histogram = [0] * 16
```

```python
    for i in range(len(chain_code) - 1):
        dx = chain_code[i + 1][0][0] - chain_code[i][0][0]
        dy = chain_code[i + 1][0][1] - chain_code[i][0][1]
        distance = int(math.sqrt(dx ** 2 + dy ** 2) * 15 / max_distance) if max_distance != 0 else 0
        distance_histogram[distance] += 1

    # 打印距离直方图
    for i, count in enumerate(distance_histogram):
        print(f'Distance {i}: {count}')

    # 显示轮廓和特征提取结果
    cv2.drawContours(image, [longest_contour], 0, (0, 0, 255), 2)
    cv2.imshow('Contour', image)
    cv2.waitKey(0)
    cv2.destroyAllWindows()
```

在上述代码中，首先读取图像并将其转换为灰度图像，对灰度图像进行二值化处理，通过阈值将图像转换为黑白两色。然后使用 cv2.findContours()函数找到图像中的轮廓，根据轮廓的面积排序，选择最大的轮廓作为感兴趣的轮廓。接下来，使用 cv2.approxPolyDP()函数对感兴趣的轮廓进行多边形逼近，得到轮廓的链码表示。最后，计算链码中每个相邻点的距离，统计距离的分布，并绘制距离分布直方图，展示轮廓的形状特征。代码执行后的效果如图 4-3 所示。

图 4-3　使用轮廓近似方法实现链码特征提取的效果

2．形状上下文

形状上下文（shape context）是一种常用的形状特征提取方法，它通过描述物体轮廓上的点与其他点之间的关系来表示形状信息。形状上下文方法基于以下两个关键思想：

- 形状点的位置不足以完整描述形状，需要考虑它与其他点的相对位置关系。
- 形状上下文特征用来描述形状点与其他点之间的相对距离和角度。

形状上下文的提取步骤如下：

（1）选择一组形状点（例如轮廓上的点）作为参考点集。
（2）计算每个参考点与其他点之间的相对距离和角度。
（3）将这些距离和角度信息组成一个向量，形成形状上下文特征。

在 Python 中可以使用 mahotas 库实现形状特征提取功能，它提供了一个名为 mahotas.features.zernike_moments 的函数，用于计算 zernike 矩特征。例如，下面的代码演示了这一用法。

实例 4-11：使用库 mahotas 实现形状特征提取

源码路径： daima\4\shangxia.py

```python
import cv2
import numpy as np
import mahotas.features

# 读取图像
image = cv2.imread('shape.jpg', cv2.IMREAD_GRAYSCALE)
```

```
# 二值化图像
_, threshold = cv2.threshold(image, 127, 255, cv2.THRESH_BINARY)

# 寻找轮廓
contours, _ = cv2.findContours(threshold, cv2.RETR_EXTERNAL, cv2.CHAIN_APPROX_NONE)

# 提取第一个轮廓
contour = contours[0]

# 计算 Zernike 矩特征
zernike_moments = mahotas.features.zernike_moments(contour, radius=21)

# 打印特征向量
print(zernike_moments)
```

在上述代码中，使用 OpenCV 库读取图像，然后进行二值化处理。接下来，使用 OpenCV 的 findContours 函数找到图像中的轮廓，并选择第一个轮廓。最后，使用 mahotas 库的 zernike_moments 函数计算轮廓的 zernike 矩特征。

4.4.3 模型拟合方法

模型拟合方法假设物体的形状可以由特定的数学模型来表示，然后通过对模型参数进行拟合来提取形状特征。常用的方法有：

1．椭圆拟合

椭圆拟合（ellipse fitting）将物体边界拟合为椭圆，并提取椭圆参数作为形状特征。当使用椭圆拟合进行图像特征提取时，通常的步骤如下：

（1）读取图像并进行预处理，例如灰度化、二值化等操作。
（2）检测图像中的轮廓，可以使用图像处理库（如 OpenCV）中的函数进行轮廓检测。
（3）对每个轮廓应用椭圆拟合算法，以获得拟合的椭圆参数。
（4）根据椭圆参数提取特征，例如椭圆的中心坐标、长轴长度、短轴长度、旋转角度等。
例如，下面是一个使用 OpenCV 库进行椭圆拟合的例子。

实例 4-12：使用 OpenCV 库进行椭圆拟合

源码路径：daima\4\tuo.py

```
import cv2
import numpy as np

# 读取图像并进行预处理
image = cv2.imread('image.jpg')
gray = cv2.cvtColor(image, cv2.COLOR_BGR2GRAY)
_, thresh = cv2.threshold(gray, 127, 255, cv2.THRESH_BINARY)

# 轮廓检测
contours, _ = cv2.findContours(thresh, cv2.RETR_EXTERNAL, cv2.CHAIN_APPROX_SIMPLE)

# 对每个轮廓应用椭圆拟合
ellipses = []
for contour in contours:
    if len(contour) >= 5:
```

```
            ellipse = cv2.fitEllipse(contour)
            ellipses.append(ellipse)

# 提取特征
for ellipse in ellipses:
    center, axes, angle = ellipse
    x, y = map(int, center)
    major_axis, minor_axis = map(int, axes)
    rotation_angle = int(angle)

    # 在图像上绘制椭圆
    cv2.ellipse(image, ellipse, (0, 255, 0), 2)

    # 打印特征信息
    print("椭圆中心坐标: ", (x, y))
    print("长轴长度: ", major_axis)
    print("短轴长度: ", minor_axis)
    print("旋转角度: ", rotation_angle)

# 显示结果图像
cv2.imshow("Ellipse Fitting", image)
cv2.waitKey(0)
cv2.destroyAllWindows()
```

在上述代码中，首先读取图像并进行预处理，然后使用 cv2.findContours()函数检测图像中的轮廓。接下来，对每个轮廓应用 cv2.fitEllipse()函数进行椭圆拟合，获取拟合的椭圆参数。最后，根据椭圆参数提取特征并在图像上绘制椭圆。代码执行后的效果如图 4-4 所示。

注意，该实例仅演示了基本的椭圆拟合和特征提取过程。根据实际需求，可能需要根据拟合结果进行更复杂的特征提取和分析。

2．直线拟合

直线拟合（line fitting）是将物体边界拟合为直线段，并提取直线参数作为形状特征。当使用直线拟合进行图像特征提取时，基本实现步骤如下：

（1）读取图像并进行预处理，例如灰度化、二值化等操作。

（2）检测图像中的边缘，可以使用边缘检测算法（如 canny 边缘检测）。

（3）根据边缘图像，检测图像中的直线段。

（4）对检测到的直线段应用直线拟合算法，以获得拟合的直线参数。

图 4-4　使用 OpenCV 库进行椭圆拟合的效果

（5）根据直线参数提取特征，例如直线的斜率、截距等。

例如，下面是一个使用 OpenCV 库实现直线拟合的例子。

实例 4-13：使用 OpenCV 库实现直线拟合

源码路径： daima\4\zhi.py

```
import cv2
import numpy as np
```

```python
# 读取图像并进行预处理
image = cv2.imread('image.jpg')
gray = cv2.cvtColor(image, cv2.COLOR_BGR2GRAY)
edges = cv2.Canny(gray, 50, 150)

# 检测直线段
lines = cv2.HoughLinesP(edges, 1, np.pi / 180, threshold=100, minLineLength=100, maxLineGap=10)

# 绘制检测到的直线
for line in lines:
    x1, y1, x2, y2 = line[0]
    cv2.line(image, (x1, y1), (x2, y2), (0, 255, 0), 2)

# 提取特征
for line in lines:
    x1, y1, x2, y2 = line[0]

    # 计算直线斜率和截距
    slope = (y2 - y1) / (x2 - x1)
    intercept = y1 - slope * x1

    # 打印特征信息
    print("直线斜率: ", slope)
    print("直线截距: ", intercept)

# 显示结果图像
cv2.imshow("Line Fitting", image)
cv2.waitKey(0)
cv2.destroyAllWindows()
```

在上述代码中，首先读取图像并进行预处理，然后使用 canny 边缘检测算法获取图像的边缘图像。接下来，使用 cv2.HoughLinesP() 函数检测图像中的直线段。根据检测到的直线段在图像上绘制直线。最后，根据直线参数提取特征，例如直线的斜率和截距。代码执行后的效果如图 4-5 所示。

图 4-5　使用 OpenCV 库实现直线拟合的效果

4.4.4 形状上的变换

形状上的变换方法通过将形状进行变换（如缩放、旋转和平移等）来提取形状特征。常用的方法有：

1. 尺度不变特征变换

尺度不变特征变换（scale-invariant feature transform，SIFT）是通过对形状进行尺度空间的变换，提取尺度不变的形状特征。尺度不变特征变换是一种用于图像特征提取的算法，它可以在不同尺度、旋转和光照条件下检测和描述图像中的关键点。SIFT 算法具有良好的尺度不变性和旋转不变性，因此在目标识别、图像匹配和三维重建等领域得到广泛应用。SIFT 算法的主要步骤如下：

（1）尺度空间构建。通过使用高斯差分函数对图像进行多次平滑和差分操作，构建尺度空间。

（2）关键点检测。在尺度空间中寻找极值点，作为关键点候选。

（3）关键点定位。通过在尺度空间的极值点周围进行精确定位，排除低对比度和边缘响应不明显的关键点。

（4）方向分配。为每个关键点分配主方向，用于后续的旋转不变性。

● 关键点描述。基于关键点周围的图像区域计算特征向量，形成关键点的描述子。

● 特征匹配。通过计算描述子之间的相似性，实现关键点的匹配。

例如，下面是使用 OpenCV 库实现 SIFT 算法的例子。

实例 4-14：使用 OpenCV 库实现 SIFT 算法

源码路径：daima\4\chidu.py

```
import cv2

# 读取图像
image = cv2.imread('image.jpg')

# 创建SIFT对象
sift = cv2.SIFT_create()

# 检测关键点和计算描述子
keypoints, descriptors = sift.detectAndCompute(image, None)

# 绘制关键点
image_with_keypoints = cv2.drawKeypoints(image, keypoints, None)

# 显示结果图像
cv2.imshow("SIFT Features", image_with_keypoints)
cv2.waitKey(0)
cv2.destroyAllWindows()
```

本实例展示了如何使用 SIFT 算法提取图像的关键点和描述子，并将关键点绘制在图像上。SIFT 算法具有较强的尺度不变性和旋转不变性，因此提取的特征可以在不同尺度和旋转条件下进行匹配和识别。在上述代码中，首先读取图像，并使用函数 cv2.SIFT_create()创建 SIFT 对象。然后，使用 sift.detectAndCompute() 函数检测关键点并计算描述子。接下来，使用 cv2.drawKeypoints()函数将关键点绘制在图像上。最后，显示结果图像。代码执行后的效果如

图 4-6 所示。

注意：SIFT 算法是一种经典的特征提取方法，但由于其涉及到专利问题，OpenCV 的最新版本中可能没有 SIFT 算法。可以使用之前的版本，或者考虑其他开源实现的 SIFT 算法，如 VLFeat 库等。

2．主成分分析

主成分分析（principal component analysis，PCA）是通过对形状进行主成分分析，提取主要的形状特征。主成分分析是一种常用的降维技术，也可以用于图像形状特征提取。PCA 可以通过线性变换将原始数据转换为新的坐标系，使得数据在新坐标系下具有最大的方差。在图像形状特征提取中，PCA 可以帮助我们找到最具代表性的形状特征，从而实现形状的描述和分类。使用 PCA 进行形状特征提取的基本步骤如下：

图 4-6　使用 OpenCV 库实现 SIFT 算法的效果

（1）数据准备。将形状数据表示为一组特征向量或特征点集合的形式。每个特征向量或特征点表示形状的一部分。

（2）特征标准化。对特征进行标准化处理，使其具有相同的尺度和范围。可以使用均值移除和缩放等方法来实现标准化。

（3）协方差矩阵计算。计算特征的协方差矩阵，表示不同特征之间的相关性。

（4）特征值分解。对协方差矩阵进行特征值分解，得到特征值和特征向量。

（5）特征选择。选择具有最大特征值的特征向量，这些特征向量对应的特征是数据中最具代表性的形状特征。

（6）特征投影。将原始数据投影到选定的特征向量上，得到新的特征表示，即形状特征。

例如，下面是一个使用 scikit-learn 库实现 PCA 形状特征提取的例子。

实例 4-15：使用 scikit-learn 库实现 PCA 形状特征提取

源码路径： daima\4\zhu.py

```python
import numpy as np
from sklearn.decomposition import PCA

# 假设有一个形状数据集，表示为特征向量的集合
shape_data = np.array([[1, 2, 3],
                       [4, 5, 6],
                       [7, 8, 9],
                       [10, 11, 12]])

# 创建 PCA 对象
pca = PCA(n_components=2)

# 执行 PCA 降维
shape_features = pca.fit_transform(shape_data)

# 输出降维后的形状特征
print(shape_features)
```

在上述代码中，首先准备了一个形状数据集，表示为一个特征向量的集合。然后，创建了

一个 PCA 对象，并指定要保留的主成分数量为 2。接下来，使用 fit_transform()方法执行 PCA 降维操作，并将原始形状数据集转换为降维后的形状特征表示。最后，打印输出降维后的形状特征。执行后会输出：

```
[[ 7.79422863  0.        ]
 [ 2.59807621  0.        ]
 [-2.59807621  0.        ]
 [-7.79422863 -0.        ]]
```

本实例展示了如何使用 PCA 进行形状特征提取。通过选择适当的主成分数量，我们可以获得最具代表性的形状特征，从而实现形状的描述和分类。

注意：PCA 是一种常见的降维技术，也可以用于形状特征提取。除了上述示例中的简单数据集，PCA 还可以应用于更复杂的形状数据，例如图像的轮廓或特征点集合。在实际应用中，可以根据具体的问题和数据类型选择适当的形状表示方法和 PCA 参数设置。

本节介绍的方法只是一些常用的形状特征提取方法，在实际应用中还可以根据具体任务和数据特点选择适合的方法。形状特征的选择和提取需要结合具体的应用场景和需求，并考虑图像的噪声、变形、光照等因素。

4.5 基于 LoG、DoG 和 DoH 的斑点检测器

斑点检测器是一种常用的图像处理技术，用于检测图像中的离散点、小斑点或孤立的亮暗区域。在斑点检测中，LoG（laplacian of gaussian）、DoG（difference of gaussian）和 DoH（determinant of hessian）是常用的滤波器或特征算子。

4.5.1 LoG

LoG 是一种线性滤波器，它是将高斯滤波器应用于图像之后再计算拉普拉斯算子。这个过程可以通过以下步骤实现：

（1）在不同尺度下应用高斯滤波器，通过改变滤波器的标准差来改变尺度。
（2）对每个尺度下的滤波结果计算拉普拉斯算子，可以通过二阶导数近似实现。
（3）在每个尺度上检测局部极值点，即图像中的斑点。
（4）LoG 滤波器的优点是可以通过不同的尺度对斑点进行多尺度检测，从而获得不同尺寸的斑点。

当使用 Python 进行图像处理时，可以使用 OpenCV 库来实现 LoG 斑点检测器进行图像特征提取。例如，下面是一个使用 OpenCV 库实现 LoG 斑点检测器的例子。

实例 4-16：使用 OpenCV 库实现 LoG 斑点检测器

源码路径：daima\4\log.py

```
import cv2

# 读取图像
image = cv2.imread('image.jpg', cv2.IMREAD_GRAYSCALE)

# 创建 LoG 滤波器
log_filter = cv2.Laplacian(image, cv2.CV_64F)

# 检测局部极值点
```

```
    keypoints = []
    threshold = 0.01  # 设定阈值
    for i in range(1, log_filter.shape[0]-1):
        for j in range(1, log_filter.shape[1]-1):
            neighbors = log_filter[i-1:i+2, j-1:j+2].flatten()
            max_neighbour = max(neighbors)
            min_neighbour = min(neighbors)
            if log_filter[i, j] > threshold and (log_filter[i, j] > max_neighbour or
log_filter[i, j] < min_neighbour):
                keypoints.append(cv2.KeyPoint(j, i, _size=2))  # 将检测到的点添加到关键点列表中

    # 在图像上绘制关键点
    output_image = cv2.drawKeypoints(image, keypoints, None, color=(0, 0, 255),
flags=cv2.DRAW_MATCHES_FLAGS_DRAW_RICH_KEYPOINTS)

    # 显示结果图像
    cv2.imshow('LoG Feature Detection', output_image)
    cv2.waitKey(0)
    cv2.destroyAllWindows()
```

在上述代码中，首先读取了一张灰度图像，然后使用 cv2.Laplacian()函数创建了 LoG 滤波器。接下来，通过遍历滤波器的每个像素，检测局部极值点，并将其添加到关键点列表中。最后，使用 cv2.drawKeypoints()函数将关键点绘制在原始图像上，并显示结果图像。注意，阈值的选择会对结果产生影响，可以根据实际情况进行调整。此外，还可以使用关键点描述算法（如 SIFT、SURF 等）对检测到的关键点进行进一步描述和匹配。代码执行后的效果如图 4-7 所示。

图 4-7　使用 OpenCV 库实现 LoG 斑点检测器的效果

4.5.2　DoG

DoG 是一种非线性滤波器，它是通过计算两个不同尺度的高斯滤波器之间的差异来实现的。DoG 滤波器的计算过程如下：

（1）在不同尺度下应用两个高斯滤波器，分别具有不同的标准差。
（2）对两个滤波结果进行相减得到 DoG 图像。
（3）在 DoG 图像中检测局部极值点（即图像中的斑点）。
（4）DoG 滤波器的优点是可以通过调整两个高斯滤波器的标准差来控制斑点的尺度。
例如，下面是一个实现 DoG 斑点检测器的例子。

实例 4-17：使用 Python 实现 DoG 斑点检测器

源码路径： daima\4\dog.py

```
import cv2
import numpy as np
```

```python
# 读取图像
image = cv2.imread('image.jpg', cv2.IMREAD_GRAYSCALE)

# 创建DoG滤波器
sigma1 = 1.0  # 第一个高斯滤波器的标准差
sigma2 = 1.6  # 第二个高斯滤波器的标准差
k = np.sqrt(2)  # 尺度因子
s1 = int(2 * np.ceil(3 * sigma1) + 1)  # 第一个高斯滤波器的大小
s2 = int(2 * np.ceil(3 * sigma2) + 1)  # 第二个高斯滤波器的大小
gaussian1 = cv2.GaussianBlur(image, (s1, s1), sigma1)
gaussian2 = cv2.GaussianBlur(image, (s2, s2), sigma2)
dog_filter = gaussian1 - k * gaussian2

# 检测局部极值点
keypoints = []
threshold = 0.01  # 设定阈值
for i in range(1, dog_filter.shape[0]-1):
    for j in range(1, dog_filter.shape[1]-1):
        neighbors = dog_filter[i-1:i+2, j-1:j+2].flatten()
        max_neighbour = max(neighbors)
        min_neighbour = min(neighbors)
        if dog_filter[i, j] > threshold and (dog_filter[i, j] > max_neighbour or dog_filter[i, j] < min_neighbour):
            keypoints.append(cv2.KeyPoint(j, i, _size=2))  # 将检测到的点添加到关键点列表中

# 在图像上绘制关键点
output_image = cv2.drawKeypoints(image, keypoints, None, color=(0, 0, 255), flags=cv2.DRAW_MATCHES_FLAGS_DRAW_RICH_KEYPOINTS)

# 显示结果图像
cv2.imshow('DoG Feature Detection', output_image)
cv2.waitKey(0)
cv2.destroyAllWindows()
```

在上述代码中，首先读取了一张灰度图像。然后，根据给定的参数（标准差、尺度因子），通过调用 cv2.GaussianBlur() 函数创建了两个不同尺度的高斯滤波器，并对图像进行滤波操作。接下来，计算 DoG 滤波器，即两个高斯滤波器之间的差异。然后，通过遍历 DoG 滤波器的每个像素，检测局部极值点，并将其添加到关键点列表中。最后，使用 cv2.drawKeypoints() 函数将关键点绘制在原始图像上，并显示结果图像。与前面的 LoG 例子类似，可以根据实际情况调整阈值和参数的值，以获得最佳的特征提取结果。同样，还可以使用关键点描述算法（如 SIFT、SURF 等）对检测到的关键点进行进一步描述和匹配。代码执行后的效果如图 4-8 所示。

图 4-8　DoG 斑点检测器的效果

4.5.3　DoH

DoH 是一种基于 Hessian 矩阵的特征检测方法，通过计算 Hessian 矩阵的行列式来检测图像中的斑点。DoH 的计算过程如下：

（1）对图像进行高斯滤波，通过改变滤波器的标准差来改变尺度。

（2）在每个尺度上计算图像的 Hessian 矩阵，包括二阶导数的信息。

（3）计算 Hessian 矩阵的行列式，并检测行列式的局部极值点（即图像中的斑点）。
（4）DoH 算法的优点是可以检测不同尺度和不同方向上的斑点。

例如，下面是一个使用 Python 实现 DoH 斑点检测器的例子。

实例 4-18：使用 Python 实现 DoH 斑点检测器

源码路径：daima\4\doh.py

```python
import cv2
import numpy as np

# 读取图像
image = cv2.imread('image.jpg', cv2.IMREAD_GRAYSCALE)

# 创建 DoH 滤波器
sigma = 1.0  # 高斯滤波器的标准差
s = int(2 * np.ceil(3 * sigma) + 1)  # 高斯滤波器的大小
gaussian = cv2.GaussianBlur(image, (s, s), sigma)
doh_filter = cv2.Laplacian(gaussian, cv2.CV_64F)

# 检测局部极值点
keypoints = []
threshold = 0.01  # 设定阈值
for i in range(1, doh_filter.shape[0]-1):
    for j in range(1, doh_filter.shape[1]-1):
        neighbors = doh_filter[i-1:i+2, j-1:j+2].flatten()
        max_neighbour = max(neighbors)
        min_neighbour = min(neighbors)
        if doh_filter[i, j] > threshold and (doh_filter[i, j] > max_neighbour or doh_filter[i, j] < min_neighbour):
            keypoints.append(cv2.KeyPoint(j, i, _size=2))  # 将检测到的点添加到关键点列表中

# 在图像上绘制关键点
output_image = cv2.drawKeypoints(image, keypoints, None, color=(0, 0, 255), flags=cv2.DRAW_MATCHES_FLAGS_DRAW_RICH_KEYPOINTS)

# 显示结果图像
cv2.imshow('DoH Feature Detection', output_image)
cv2.waitKey(0)
cv2.destroyAllWindows()
```

在上述代码中，首先读取了一张灰度图像。然后，根据给定的标准差，通过调用 cv2.GaussianBlur()函数创建了高斯滤波器，并对图像进行滤波操作。接下来，使用 cv2.Laplacian()函数计算 Hessian 矩阵的行列式，得到 DoH 滤波器。然后，通过遍历 DoH 滤波器的每个像素，检测局部极值点，并将其添加到关键点列表中。最后，使用 cv2.drawKeypoints()函数将关键点绘制在原始图像上，并显示结果图像。代码执行后的效果如图 4-9 所示。

图 4-9　DoH 斑点检测器的效果

注意：在本节中介绍的斑点检测器，大家在实际应用中可以根据需求选择使用。LoG 和 DoG 通常用于多尺度斑点检测，可以获得不同尺寸的斑点。而 DoH 则可以更准确地检测具有不同尺度和方向的斑点。根据图像特点和应用需求，选择适合的斑点检测器可以提高检测效果和准确性。

第 5 章　图像分割处理

图像分割是一种计算机视觉任务，旨在将图像中的像素分成不同的语义类别或物体实例。其目标是根据图像中像素的特征和上下文信息，将图像分割为具有特定语义意义的区域。图像分割在许多应用领域中发挥着重要作用，包括医学影像分析、自动驾驶、图像编辑和虚拟现实等。本章详细讲解使用 Python 语言实现图像分割的知识。

5.1　图像分割的重要性

图像分割是将图像中的像素划分为不同的类别或实例的过程，可以通过各种方法实现，包括基于阈值、边缘、区域、图论和深度学习等。可以根据具体的应用场景和需求选择适合的方法。图像分割在计算机视觉领域中非常重要，下面简要介绍图像分割的几个重要性：

- 目标检测和识别：图像分割可以帮助我们在图像中准确地定位和分割出特定的目标或物体。这对于目标检测和识别任务至关重要，例如自动驾驶中的车辆和行人检测，医学影像中的病灶定位等。通过将图像分割成目标区域，可以提供更准确的目标定位和边界信息，有助于后续的分析和处理。
- 图像理解和语义分析：图像分割可以提供对图像内容更详细的理解和分析。通过将图像分割为语义区域，可以获取每个区域的特征和上下文信息。这对于图像理解、场景分析、图像注释等任务非常有用。例如，在自然图像中分割出不同的物体和场景可以帮助计算机理解图像的语义含义。
- 图像编辑和合成：图像分割是图像编辑和合成的重要工具。通过对图像进行分割，可以对不同区域进行独立的编辑和处理，例如去除或替换特定物体、修改背景、图像合成等。这对于图像处理、图形设计和虚拟现实等应用具有重要意义。
- 医学影像和生物图像分析：在医学影像和生物图像领域，图像分割对于病灶定位、器官分割和形状分析非常重要。它可以帮助医生和研究人员在影像数据中精确地提取感兴趣的区域，从而辅助诊断和治疗决策。

总的来说，图像分割在许多计算机视觉任务和应用中都起着关键作用。它提供了对图像内容的细粒度理解和处理能力，有助于实现更准确、高效和智能的图像分析和应用。实现图像分割的主要方法有：基于阈值的分割、基于边缘的分割、基于区域的分割、基于图论的分割、基于深度学习的分割。

5.2　基于阈值的分割

基于阈值的图像分割是一种简单且常用的分割方法，它基于像素的灰度值或颜色信息设置一个或多个阈值，将图像分成不同的区域。这种方法适用于图像中具有明显不同颜色或灰度级别的区域。

5.2.1 灰度阈值分割

灰度阈值分割是将灰度图像根据像素的灰度值进行分割的方法。通过选择一个合适的灰度阈值，将图像分成两个区域：高于阈值的像素归为一类，低于阈值的像素归为另一类。例如，下面是一个使用 OpenCV 库实现灰度阈值分割的例子。

实例 5-1：使用 OpenCV 库实现灰度阈值分割

源码路径：daima\5\hui.py

```python
import cv2

# 读取灰度图像
image = cv2.imread('888.jpg', 0)

# 应用阈值分割
threshold_value = 128
_, segmented_image = cv2.threshold(image, threshold_value, 255, cv2.THRESH_BINARY)

# 显示原始图像和分割后的图像
cv2.imshow('Original Image', image)
cv2.imshow('Segmented Image', segmented_image)
cv2.waitKey(0)
cv2.destroyAllWindows()
```

在上述代码中，首先使用 cv2.imread()函数读取一幅灰度图像。然后，使用 cv2.threshold()函数应用阈值分割。函数的参数包括图像、阈值、最大像素值（这里设为 255，表示分割后的前景像素为白色），以及分割方法（这里使用二值化分割）。函数的返回值包括阈值和分割后的图像。最后，使用 cv2.imshow()函数显示原始图像和分割后的图像，使用 cv2.waitKey() 等待按键操作，使用 cv2.destroyAllWindows()关闭窗口。代码执行后的效果如图 5-1 所示。

图 5-1　使用 OpenCV 库实现的灰度阈值分割效果

5.2.2 彩色阈值分割

彩色阈值分割是将彩色图像根据像素的颜色信息进行分割的方法。通过选择一个或多个颜色阈值，将图像中的不同颜色区域分割开来。在彩色图像中，每个像素由 RGB（红、绿、蓝）三个通道的值组成。通过选择适当的颜色阈值，可以将图像中的不同颜色区域分割开来。例如，下面是一个使用 OpenCV 库实现彩色阈值分割的例子。

实例 5-2：使用 OpenCV 库实现彩色阈值分割

源码路径：daima\5\cai.py

```python
import cv2
```

```python
import numpy as np

# 读取彩色图像
image = cv2.imread('888.jpg')

# 将图像转换为HSV颜色空间
hsv_image = cv2.cvtColor(image, cv2.COLOR_BGR2HSV)

# 定义颜色阈值范围（例如，红色苹果）
lower_threshold = np.array([0, 50, 50])    # 最低阈值
upper_threshold = np.array([10, 255, 255]) # 最高阈值

# 创建掩模，将位于阈值范围内的像素设置为白色（255），位于范围外的像素设置为黑色（0）
mask = cv2.inRange(hsv_image, lower_threshold, upper_threshold)

# 对原始图像和掩模应用位操作，提取分割后的图像
segmented_image = cv2.bitwise_and(image, image, mask=mask)

# 显示原始图像和分割后的图像
cv2.imshow('Original Image', image)
cv2.imshow('Segmented Image', segmented_image)
cv2.waitKey(0)
cv2.destroyAllWindows()
```

在上述代码中，首先使用 cv2.imread()函数读取一幅彩色图像。然后，将图像从 BGR 颜色空间转换为 HSV 颜色空间，通过 cv2.cvtColor()函数实现。这是因为在 HSV 颜色空间中，颜色可以更容易地表示为阈值范围。接下来定义了颜色阈值范围，以提取感兴趣的颜色区域。在这个例子中，选择提取红色，因此定义了红色的阈值范围。通过 np.array()函数创建了一个包含最低阈值和最高阈值的 NumPy 数组。然后，使用 cv2.inRange()函数创建了一个掩模（mask），将位于阈值范围内的像素设置为白色（255），位于范围外的像素设置为黑色（0）。使用 cv2.bitwise_and()函数对原始图像和掩模进行位操作，提取分割后的图像。使用 cv2.imshow()函数显示原始图像和分割后的图像，再使用 cv2.waitKey()函数等待按键操作，最后使用 cv2.destroyAllWindows()函数关闭窗口。

本实例演示了如何使用彩色阈值分割方法提取感兴趣的颜色区域，大家可以根据需要调整阈值范围和颜色选择来实现不同的分割效果。

5.3 基于边缘的分割

基于边缘的图像分割是一种常用的分割方法，它基于图像中物体边缘的特征来进行分割。边缘是图像中亮度或颜色发生剧烈变化的区域，通常表示不同物体或物体的边界。基于边缘的分割方法通过检测和连接边缘来实现图像分割。基于边缘的图像分割方法在许多应用中都非常有用，特别是对于提取物体边界和形状信息。它们在计算机视觉、图像处理和模式识别等领域得到广泛应用。

5.3.1 canny 边缘检测

canny 边缘检测是一种经典的边缘检测算法，它在图像中寻找强度梯度最大的位置，并将其视为边缘点。实现 canny 边缘检测的基本步骤如下：

(1)将图像转换为灰度图像。
(2)对灰度图像应用高斯滤波来平滑图像,减少噪声。
(3)计算图像中每个像素的梯度幅值和方向。
(4)应用非极大值抑制,将非边缘像素抑制为 0,保留梯度幅值最大的边缘像素。
(5)应用双阈值处理,通过设置高阈值和低阈值来确定强边缘和弱边缘,并将它们连接成完整的边缘。

canny 边缘检测使用了多个步骤来实现边缘检测,包括高斯滤波、梯度计算、非极大值抑制和双阈值处理。例如,下面是一个使用 OpenCV 库实现 canny 边缘检测的例子。

实例 5-3:使用 OpenCV 库实现 canny 边缘检测

源码路径:daima\5\can.py

```
import cv2

# 读取图像
image = cv2.imread('888.jpg', 0)

# 应用 canny 边缘检测
edges = cv2.Canny(image, 100, 200)    # 阈值 1 为 100,阈值 2 为 200

# 显示原始图像和边缘图像
cv2.imshow('Original Image', image)
cv2.imshow('Edges', edges)
cv2.waitKey(0)
cv2.destroyAllWindows()
```

在上述代码中,首先使用 cv2.imread()函数读取一幅灰度图像。然后,使用 cv2.Canny()函数应用 canny 边缘检测。函数的参数包括图像、两个阈值,阈值 1 用于边缘的强度梯度下限,阈值 2 用于边缘的强度梯度上限。接下来,使用 cv2.imshow()函数显示原始图像和边缘图像,使用 cv2.waitKey()等待按键操作,最后使用 cv2.destroy-AllWindows()关闭窗口。本实例演示了如何使用 canny 边缘检测算法实现图像分割,可以根据需要调整阈值的选择,以获得更好的边缘检测效果。代码执行后的效果如图 5-2 所示。

图 5-2 使用 OpenCV 库实现 canny 边缘检测的效果

5.3.2 边缘连接方法

基于边缘连接的图像分割方法是一种基于边缘像素的相邻性来连接边缘的方法。这种方法旨在将图像中的边缘像素连接成连续的边缘,从而实现图像的分割,形成完整的边缘。这些方法包括:

● 滞后阈值:根据梯度幅值的阈值,将边缘像素标记为强边缘或弱边缘。然后,根据弱边缘像素是否与强边缘像素相连,决定将其保留还是抑制。

- 连接分析：通过在边缘像素之间建立连接关系，将相邻的边缘像素连接起来形成连续的边缘。
- 边缘跟踪：从一个初始边缘像素开始，沿着边缘的方向追踪相邻的边缘像素，直到边缘结束。

例如，下面是一个使用 OpenCV 库实现基于边缘连接方法的图像分割例子。

实例 5-4：使用 OpenCV 库实现基于边缘连接方法的图像分割
源码路径： daima\5\bian.py

```
import cv2
import numpy as np

# 读取图像并进行边缘检测
image = cv2.imread('888.jpg', 0)
edges = cv2.Canny(image, 100, 200)

# 执行边缘连接方法
contours, _ = cv2.findContours(edges, cv2.RETR_EXTERNAL, cv2.CHAIN_APPROX_SIMPLE)

# 创建一个黑色背景图像
segmented_image = np.zeros_like(image)

# 在黑色背景上绘制边缘
cv2.drawContours(segmented_image, contours, -1, (255, 255, 255), thickness=cv2.FILLED)

# 显示原始图像和分割后的图像
cv2.imshow('Original Image', image)
cv2.imshow('Segmented Image', segmented_image)
cv2.waitKey(0)
cv2.destroyAllWindows()
```

对上述代码的具体说明如下：

（1）首先使用函数 cv2.imread()读取一幅灰度图像。再使用函数 cv2.Canny()对图像进行边缘检测，生成一组离散的边缘像素。

（2）然后，使用函数 cv2.findContours()执行边缘连接方法，从边缘图像中找到边缘的连续轮廓。参数 cv2.RETR_EXTERNAL 表示只检测最外层的边缘，cv2.CHAIN_APPROX_SIMPLE 表示仅保留边缘像素的端点。

（3）接下来，创建一个与原始图像大小相同的黑色背景图像作为分割后的图像。使用函数 cv2.drawContours()在黑色背景上绘制边缘轮廓，将边缘像素连接起来形成连续的边缘。

（4）最后，使用函数 cv2.imshow()显示原始图像和分割后的图像，然后使用函数 cv2.waitKey()等待按键操作，使用函数 cv2.destroyAllWindows()关闭窗口。

5.4 基于区域的分割

基于区域的图像分割方法是一种常见的图像分割技术，它基于图像中不同区域的特征来进行分割。该方法将图像分成具有相似特征的区域。它通常根据像素的颜色、纹理、形状和像素之间的距离等特征进行分割。一种常见的区域分割算法是基于区域增长，它从种子像素开始，通过合并具有相似特征的相邻像素来逐步扩展区域。

5.4.1 区域生长算法

区域生长算法是一种基于像素相似性的图像分割方法。它从一个或多个种子像素开始，根据一定的准则和规则，逐渐将与种子像素相似的邻域像素加入同一区域，形成连续的区域。该算法通常包括以下步骤：

（1）选择种子像素或种子区域。

（2）定义区域生长的准则，如像素的灰度值相似性、颜色相似性等。

（3）逐个处理邻域像素，根据准则将其加入区域。

（4）重复上述步骤，直到无法再添加像素或达到停止准则。

例如，下面是一个使用区域生长算法实现图像分割的例子。

实例 5-5：使用区域生长算法实现图像分割

源码路径： daima\5\qu.py

```
import numpy as np
import cv2

# 读取图像
image = cv2.imread('888.jpg', 0)

# 区域生长算法
def region_growing(image, seed, threshold):
    height, width = image.shape
    segmented = np.zeros_like(image, dtype=np.uint8)
    visited = np.zeros_like(image, dtype=np.uint8)

    stack = []
    stack.append(seed)

    while stack:
        x, y = stack.pop()

        if visited[x, y] == 1:
            continue

        visited[x, y] = 1

        if abs(int(image[x, y]) - int(image[seed])) <= threshold:  # 设置生长准则
            segmented[x, y] = 255

            if x > 0:
                stack.append((x - 1, y))
            if x < height - 1:
                stack.append((x + 1, y))
            if y > 0:
                stack.append((x, y - 1))
            if y < width - 1:
                stack.append((x, y + 1))

    return segmented

# 设置种子点并应用区域生长算法
```

```
seed_point = (100, 100)  # 设置种子点坐标
threshold = 20  # 设置生长容差
segmented_image = region_growing(image, seed_point, threshold)

# 显示原始图像和分割后的图像
cv2.imshow('Original Image', image)
cv2.imshow('Segmented Image', segmented_image)
cv2.waitKey(0)
cv2.destroyAllWindows()
```

上述代码是一个使用OpenCV库实现的简单区域生长算法的示例，用于图像分割。下面是对代码的具体说明：

（1）通过函数cv2.imread()读取一幅灰度图像，并将其存储在image变量中。

（2）定义一个函数region_growing()，该函数实现了区域生长算法。函数接受三个参数：图像（image）、种子点（seed）和阈值（threshold）。在函数region_growing()中，首先获取图像的高度和宽度，创建一个与图像大小相同的空白图像（segmented）和一个标记图像（visited），用于记录已访问过的像素。

（3）创建一个栈数据结构（stack），将种子点添加到栈中。

（4）在循环中，从栈中取出一个像素(x, y)，并检查该像素是否已被访问过。如果已经访问过，则继续循环；否则，将其标记为已访问。然后，根据生长准则判断当前像素与种子点之间的灰度差是否小于等于阈值。如果满足条件，将当前像素标记为分割区域的一部分，并将其邻域像素添加到栈中。

（5）重复执行上一步，直到栈为空。

（6）函数返回分割后的图像（segmented）。

（7）在主函数中，设置种子点的坐标和生长容差，并将图像和种子点传递给函数region_growing()进行图像分割。分割后的图像通过函数cv2.imshow()显示出来，代码执行后的效果如图5-3所示。

图5-3　使用区域增长算法实现的效果

注意：这个区域生长算法是一种简单的实现，并且对于复杂的图像可能无法得到理想的分割结果。根据具体的应用场景，你可能需要根据图像的特点调整阈值和其他参数，以获得更好的分割效果。

5.4.2　基于图论的分割算法

基于图论的分割算法是一种基于图论的图像分割方法。它将图像表示为图的形式，其中像素作为图的顶点，像素之间的关系作为图的边。通过最小化图中顶点之间的权重，将图像分割为不同的区域。基于图论的分割算法的实现步骤如下：

（1）构建图，其中顶点表示图像的像素，边表示像素之间的关系。

（2）定义顶点和边的权重，例如基于像素的颜色差异、纹理差异等。

（3）使用图割算法（如最小割最大流算法）将图像分割为多个区域。

例如，下面是一个使用 PyMaxflow 库借助于图割算法实现图像分割的例子。

实例 5-6：使用 PyMaxflow 库借助于图割算法实现图像分割
源码路径：daima\5\tuge.py

```python
import numpy as np
import cv2
import maxflow

# 读取图像
image = cv2.imread('888.jpg')
image = cv2.cvtColor(image, cv2.COLOR_BGR2RGB)

# 定义图割算法
def graph_cut_segmentation(image):
    height, width, _ = image.shape

    # 创建图割图
    g = maxflow.Graph[float]()
    nodeids = g.add_grid_nodes((height, width))

    # 设置数据项和平滑项的权重
    data_weight = 0.5
    smooth_weight = 0.2

    # 添加数据项和平滑项的边
    g.add_grid_edges(nodeids, data_weight=data_weight, smooth_weight=smooth_weight)

    # 设置种子点
    seed1 = (50, 50)
    seed2 = (200, 200)

    # 设置种子点的标签
    g.add_tedge(nodeids[seed1], 0, 255)
    g.add_tedge(nodeids[seed2], 255, 0)

    # 运行图割算法
    g.maxflow()

    # 获取分割结果
    segmented = g.get_grid_segments(nodeids)

    # 创建分割后的图像
    segmented_image = np.zeros_like(image)
    segmented_image[segmented] = image[segmented]

    return segmented_image

# 应用图割算法进行分割
segmented_image = graph_cut_segmentation(image)

# 显示原始图像和分割后的图像
plt.subplot(1, 2, 1)
plt.imshow(image)
plt.title('Original Image')
```

```
plt.subplot(1, 2, 2)
plt.imshow(segmented_image)
plt.title('Segmented Image')

plt.tight_layout()
plt.show()
```

在上述代码中，首先使用 cv2.imread()函数读取一幅图像，并将其转换为 RGB 颜色空间。然后，定义了一个 graph_cut_segmentation()函数来实现图割算法的图像分割。在 graph_cut_segmentation()函数中，首先创建了一个图割图，并添加了网格节点。设置了数据项和平滑项的权重，并添加了对应的边。接下来设置了两个种子点，并为它们分配了标签。最后运行图割算法，并根据分割结果创建分割后的图像。在主函数中，调用 graph_cut_segmentation()函数对图像进行分割，并显示原始图像和分割后的图像。代码执行后的效果如图 5-4 所示。

图 5-4　使用库 PyMaxflow 借助于图割算法实现的效果

5.4.3　基于聚类的分割算法

基于聚类的分割算法将图像中的像素聚类到不同的区域，根据像素之间的相似性或差异性来进行分割。常用的聚类算法包括 K 均值聚类、谱聚类等。基于聚类的分割算法的实现步骤如下：

（1）将图像像素表示为特征向量。
（2）使用聚类算法将像素分组到不同的聚类中心。
（3）将每个聚类标记为一个区域。

例如，下面是一个使用库 scikit-learn 实现基于聚类的图像分割的例子。

实例 5-7：使用库 scikit-learn 实现基于聚类的图像分割

源码路径： daima\5\ju.py

```
import numpy as np
import cv2
import matplotlib.pyplot as plt
from sklearn.cluster import KMeans


# 读取图像
image = cv2.imread('888.jpg')
image = cv2.cvtColor(image, cv2.COLOR_BGR2RGB)
```

```python
# 将图像转换为一维数组
pixels = image.reshape(-1, 3)

# 执行K均值聚类
kmeans = KMeans(n_clusters=4, random_state=0)
kmeans.fit(pixels)

# 获取每个像素点的标签
labels = kmeans.labels_

# 将每个像素点的标签转换为RGB值
segmented_pixels = kmeans.cluster_centers_[labels]
segmented_image = segmented_pixels.reshape(image.shape)

# 显示原始图像和分割后的图像
plt.subplot(1, 2, 1)
plt.imshow(image)
plt.title('Original Image')

plt.subplot(1, 2, 2)
plt.imshow(segmented_image.astype(np.uint8))
plt.title('Segmented Image')

plt.tight_layout()
plt.show()
```

在上述代码中，首先使用函数 cv2.imread() 读取一幅图像，并将其转换为 RGB 颜色空间。然后，将图像的像素重新排列为一维数组，以便进行聚类操作。接下来，使用 scikit-learn 库中的 K 均值算法执行聚类操作。在这个例子中，设置聚类数为 4，但可以根据需要调整聚类数。再获取每个像素点的聚类标签，并将每个像素点的标签转换回 RGB 值。最后，将分割后的像素重新排列为图像形状，并显示原始图像和分割后的图像。代码执行后的效果如图 5-5 所示。

本实例演示了如何使用 scikit-learn 库中的 K 均值算法实现基于聚类的图像分割。通过调整聚类数和其他参数，可以获得不同的分割效果。

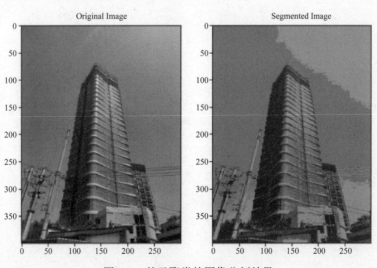

图 5-5　基于聚类的图像分割效果

5.5 最小生成树算法

最小生成树算法是基于图论的分割法中一个算法,它也可以用于图像分割。该算法通过将图像中的像素或区域表示为图的节点,并根据像素之间的相似性构建图的边权重,然后利用最小生成树算法选择边权重最小的一组边,将图分割成不同的区域。最小生成树算法通常用于基于区域的分割方法,它在保持区域的连续性和平滑性方面具有优势。

使用最小生成树算法实现图像分割的基本步骤如下:

(1) 图像表示。将图像转换为一个图的形式,其中图的节点表示图像中的像素或区域,图的边表示像素或区域之间的关系。可以使用像素或区域作为图的节点,并计算节点之间的相似性或距离作为边的权重。

(2) 构建图。基于图像的节点和边的权重构建一个无向图。节点表示图像中的像素或区域,边表示像素或区域之间的关系。边的权重可以根据像素之间的相似性、颜色差异、纹理等特征计算得出。

(3) 最小生成树。使用最小生成树算法从图中选择边,构建一个包含所有节点的最小生成树。最小生成树是一个连接图中所有节点的子图,其中边的权重之和最小。最小生成树算法有多种实现方法,包括 Prim 算法和 Kruskal 算法。

(4) 分割结果提取。根据最小生成树的结果,将图像分割成不同的区域。可以根据最小生成树中的边来确定图像中的区域边界,将图像分割成不同的区域。

例如,下面是一个在 Python 程序中使用最小生成树算法实现图像分割的例子。

实例 5-8:使用最小生成树算法实现图像分割
源码路径: daima\5\zuixiao.py

```python
import numpy as np
import matplotlib.pyplot as plt
from skimage import io, segmentation
from skimage.future import graph

# 读取图像
image = io.imread('888.jpg')

# 执行图像分割
labels = segmentation.slic(image, compactness=30, n_segments=400)
g = graph.rag_mean_color(image, labels)

# 应用最小生成树算法
labels2 = graph.cut_normalized(labels, g)

# 可视化分割结果
fig, ax = plt.subplots(nrows=1, ncols=2, figsize=(10, 5))

ax[0].imshow(image)
ax[0].set_title('Original Image')

ax[1].imshow(segmentation.mark_boundaries(image, labels2))
```

```
ax[1].set_title('Segmentation')

for a in ax:
    a.axis('off')

plt.tight_layout()
plt.show()
```

在上述代码中，首先使用函数 io.imread() 加载一张示例图像。然后，使用函数 segmentation.slic() 对图像进行超像素分割，将图像划分为多个相似的区域。这里的 compactness 参数控制了超像素的紧凑度，n_segments 参数指定了希望得到的超像素数量。接下来，使用函数 graph.rag_mean_color() 构建了一个图，其中图的节点是超像素，节点的颜色由该超像素内像素的平均颜色表示。应用最小生成树算法 graph.cut_normalized() 来将图像的超像素分割成不同的区域，这里的 labels2 变量保存了最终的分割结果。最后，使用 Matplotlib 库将原始图像和分割结果进行可视化展示。函数 segmentation.mark_boundaries() 用于在分割结果中标记出区域边界。代码执行后的效果如图 5-6 所示。

图 5-6　使用最小生成树实现的图像分割效果

5.6　基于深度学习的分割

基于深度学习的图像分割方法在近年来取得了重大的突破，成为图像分割领域的前沿技术。基于深度学习的图像分割方法利用深度神经网络模型来学习图像的语义信息，以实现对图像中不同物体或区域的准确分割。这些深度学习方法在大规模的图像分割任务中取得了显著的性能提升，同时也为图像分割在医学影像分析、自动驾驶、图像语义理解等领域的应用提供了强大的工具。要实现基于深度学习的图像分割，通常需要大量的标记数据和计算资源来训练和优化深度神经网络模型。同时，还需要了解深度学习框架（如 TensorFlow、PyTorch）和相关库的使用，以及网络架构设计、损失函数的选择等技术细节。

5.6.1 FCN（全卷积网络）

FCN 是一种经典的图像分割方法，通过将传统的卷积神经网络（CNN）结构转换为全卷积结构，使网络能够输出与输入图像具有相同尺寸的分割结果。FCN 利用卷积和反卷积操作来学习图像中每个像素的类别标签，从而实现像素级别的分割。

下面是使用 FCN 实现图像分割的基本步骤：

（1）数据准备。准备带有标记的图像数据集，其中每个图像都有对应的像素级别的标签。通常需要手动对一些图像进行标记，即为每个像素指定类别标签。

（2）构建 FCN 模型。选择合适的深度学习框架（如 TensorFlow、PyTorch）并使用该框架构建 FCN 模型。FCN 模型通常由编码器和解码器组成。编码器负责提取图像的语义信息，通常使用卷积神经网络（如 VGG、ResNet）来实现。解码器将特征图恢复到原始图像尺寸并生成分割结果，通常使用反卷积或上采样操作实现。在编码器和解码器之间，可以添加跳跃连接来融合低层和高层特征。

（3）模型训练。使用准备好的数据集对 FCN 模型进行训练。训练过程中，通过最小化损失函数来优化模型参数，使得模型能够准确地预测每个像素的类别标签。常用的损失函数包括交叉熵损失、Dice 损失等。

（4）模型评估。使用评估指标（如像素准确率、IoU）对训练好的模型进行评估，以了解模型在图像分割任务上的性能表现。

（5）图像分割。使用训练好的 FCN 模型对新的图像进行分割。将图像输入到 FCN 模型中，得到每个像素的类别标签，从而实现图像的分割。

例如，下面是一个基于 FCN 的图像分割的 Python 示例，使用的是 PyTorch 框架和预训练的 FCN 模型（FCN-8s）。

实例 5-9：使用预训练的 FCN 模型（FCN-8s）实现图像分割

```python
import torch
import torchvision
import matplotlib.pyplot as plt
from PIL import Image

# 加载预训练的 FCN 模型
fcn = torchvision.models.segmentation.fcn_resnet50(pretrained=True)

# 设置模型为评估模式
fcn.eval()

# 加载图像并进行预处理
image = Image.open('888.jpg')
preprocess = torchvision.transforms.Compose([
    torchvision.transforms.ToTensor(),
    torchvision.transforms.Normalize(mean=[0.485, 0.456, 0.406], std=[0.229, 0.224, 0.225])
])
input_tensor = preprocess(image)
input_batch = input_tensor.unsqueeze(0)

# 将输入图像传递给模型进行分割
with torch.no_grad():
    output = fcn(input_batch)['out'][0]
output_predictions = output.argmax(0)
```

```python
# 可视化分割结果
plt.figure(figsize=(10, 5))
plt.subplot(1, 2, 1)
plt.imshow(image)
plt.title('Original Image')

plt.subplot(1, 2, 2)
plt.imshow(output_predictions)
plt.title('Segmentation')

plt.show()
```

在上述代码中，需要将 your_image.jpg 替换为自己的图像文件路径。然后，通过加载预训练的 FCN 模型，对图像进行预处理，并将其输入到模型中进行分割。最后，使用 Matplotlib 库将原始图像和分割结果进行可视化。

注意：上述实例中的 FCN 模型是使用在大规模图像数据集上进行预训练的。如果读者的应用场景与预训练模型的数据集不匹配，可能需要进行微调或训练自定义的 FCN 模型，以获得更好的分割效果。

5.6.2 U-Net

U-Net 是一种常用的图像分割网络，其特点是具有 U 形的网络结构。U-Net 由编码器和解码器组成，编码器用于提取图像的语义信息，解码器用于将特征图恢复到原始图像尺寸并生成分割结果。U-Net 还引入了跳跃连接，可以将低层特征与高层特征相融合，提高分割结果的准确性。

5.6.3 DeepLab

DeepLab 是一种基于空洞卷积（dilated convolution）的图像分割方法。空洞卷积可以扩大卷积核的感受野，从而捕捉更大范围的上下文信息。DeepLab 还引入了空间金字塔池化（spatial pyramid pooling）模块，可以对不同尺度的特征进行池化和融合，提高分割结果的准确性。例如，下面是一个使用预训练的 U-Net 模型实现图像分割的例子。

实例 5-10：使用预训练的 U-Net 模型实现图像分割
源码路径：daima\5\unet.py

```python
import torch
import torchvision
import matplotlib.pyplot as plt
from PIL import Image

# 加载预训练的 U-Net 模型
unet = torchvision.models.segmentation.deeplabv3_resnet50(pretrained=True)

# 设置模型为评估模式
unet.eval()

# 加载图像并进行预处理
image = Image.open('your_image.jpg')
preprocess = torchvision.transforms.Compose([
    torchvision.transforms.ToTensor(),
    torchvision.transforms.Normalize(mean=[0.485, 0.456, 0.406], std=[0.229, 0.224, 0.225])
])
input_tensor = preprocess(image)
input_batch = input_tensor.unsqueeze(0)
```

```python
# 将输入图像传递给模型进行分割
with torch.no_grad():
    output = unet(input_batch)['out']
output_predictions = output.argmax(1)[0]

# 可视化分割结果
plt.figure(figsize=(10, 5))
plt.subplot(1, 2, 1)
plt.imshow(image)
plt.title('Original Image')

plt.subplot(1, 2, 2)
plt.imshow(output_predictions)
plt.title('Segmentation')

plt.show()
```

在上述代码中，首先需要将 your_image.jpg 替换为自己的图像文件路径。然后，通过加载预训练的 U-Net 模型，对图像进行预处理，并将其输入到模型中进行分割。最后，使用 Matplotlib 库将原始图像和分割结果进行可视化。

5.6.4 Mask R-CNN

基于 Mask R-CNN（mask region-based convolutional neural network）的方法是一种流行的图像分割方法，它结合了目标检测和语义分割的能力。Mask R-CNN 是一种深度学习模型，它可以同时预测图像中的对象位置和像素级的语义分割掩码。

下面是使用 Mask R-CNN 实现图像分割的基本步骤：

（1）数据准备。准备带有标注的训练数据集，包括图像和与图像中的每个对象关联的边界框和分割掩码。

（2）模型构建。构建 Mask R-CNN 模型，该模型通常由两个主要组件组成：共享的卷积神经网络（通常是用于特征提取的骨干网络，如 ResNet）和两个分支网络，一个用于目标检测（边界框预测）和一个用于语义分割（分割掩码预测）。

（3）模型训练。使用准备好的训练数据集对 Mask R-CNN 模型进行训练。训练过程通常包括对网络参数进行初始化，通过前向传播计算损失函数，然后通过反向传播更新网络参数。

（4）目标检测和分割预测。对于新的图像，使用训练好的模型进行目标检测和分割预测。首先，通过前向传播在图像中检测对象边界框。然后，在每个边界框的基础上，使用网络的分割分支预测每个对象的像素级分割掩码。

（5）后处理和可视化。对于每个对象的分割掩码，可以应用后处理步骤（如阈值处理、填充等）来提取准确的分割结果。最后，可以将分割结果可视化或应用于后续的图像分析任务。

（6）使用 Mask R-CNN 进行图像分割需要较大的计算资源和大规模的训练数据集。通常，可以使用已经在大型数据集（如 COCO）上进行预训练的模型来加快训练和预测过程。同时，可以根据特定的应用场景微调预训练模型，以获得更好的性能。

例如，下面是一个使用 Mask R-CNN 进行图像分割的例子。

实例 5-11：使用 Mask R-CNN 进行图像分割

源码路径：daima\5\mask.py

```
import torch
import torchvision
import matplotlib.pyplot as plt
```

```python
from PIL import Image

# 加载预训练的 Mask R-CNN 模型
mask_rcnn = torchvision.models.detection.maskrcnn_resnet50_fpn(pretrained=True)

# 设置模型为评估模式
mask_rcnn.eval()

# 加载图像并进行预处理
image = Image.open('your_image.jpg')
transform = torchvision.transforms.Compose([
    torchvision.transforms.ToTensor(),
    torchvision.transforms.Normalize(mean=[0.485, 0.456, 0.406], std=[0.229, 0.224, 0.225])
])
input_image = transform(image)

# 将输入图像传递给模型进行预测
with torch.no_grad():
    prediction = mask_rcnn([input_image])

# 提取预测结果
masks = prediction[0]['masks']
scores = prediction[0]['scores']
labels = prediction[0]['labels']

# 可视化预测结果
plt.figure(figsize=(10, 5))
plt.subplot(1, 2, 1)
plt.imshow(image)
plt.title('Original Image')

plt.subplot(1, 2, 2)
plt.imshow(masks[0, 0].mul(255).byte(), cmap='gray')
plt.title('Segmentation Mask')

plt.show()
```

在上述代码中，需要将 your_image.jpg 替换为自己的图像文件路径。代码加载了预训练的 Mask R-CNN 模型，并将图像进行预处理。然后，通过模型的前向传播进行预测，得到包含分割掩码、置信度分数和标签的预测结果。最后，使用 Matplotlib 库将原始图像和分割掩码进行可视化。

注意： Mask R-CNN 虽然在图像分割任务中表现出色，但由于其计算复杂性较高，对于实时应用和资源受限的环境可能不太适用。

第 6 章　目标检测处理

目标检测（object detection）是计算机视觉领域中的一个重要任务，其目标是在图像或视频中准确地定位和识别多个目标物体的位置和类别。本章详细讲解使用 Python 语言实现图像目标检测的知识。

6.1　目标检测介绍

目标检测在许多领域中具有广泛的应用，如自动驾驶、视频监控、人脸识别、物体识别等。目标检测为我们提供了准确和实时的目标识别和定位能力，为各种智能系统和应用带来了巨大的价值。

6.1.1　目标检测的步骤

目标检测需要在图像中确定目标物体的边界框（bounding box）并标识出它们的类别。通常，目标检测算法需要完成以下主要步骤：

（1）目标定位（localization）。在图像中定位目标物体的位置，通常用边界框表示目标的位置和大小。

（2）目标分类（classification）。将边界框中的目标物体进行分类，确定它们的类别。

6.1.2　目标检测的方法

目标检测算法可以分为两大类：基于传统方法的目标检测和基于深度学习的目标检测。

1．传统方法

传统方法的目标检测通常基于手工设计的特征提取器和机器学习算法，这些方法使用人工定义的特征，如 Haar 特征、HOG 特征等，结合分类器（如支持向量机、AdaBoost 等）来检测和分类目标。

2．深度学习方法

深度学习的目标检测使用深度卷积神经网络或卷积神经网络的改进模型来端到端地学习目标检测任务。深度学习方法通常包括如下两个主要组件：

● 区域提议网络（region proposal network，RPN）：用于生成候选目标边界框。

● 目标分类网络：用于对候选边界框进行分类和定位。

流行的深度学习目标检测算法包括 Faster R-CNN、YOLO（you only look once）、SSD（single shot multibox detector）等。

6.2　YOLO v5

YOLO 是一种基于深度学习的目标检测算法，它具有实时性和高准确性的特点。YOLO 算法通过单个神经网络同时完成目标定位和分类，以极高的速度在图像或视频中检测多个目标物体。YOLO v5 是 YOLO 算法的主流版本之一，它在原始的 YOLO 算法基础上进行了一些关键性的改进，以提升检测性能和准确性。

6.2.1　YOLO v5 的改进

YOLO v5 是 YOLO 算法的改进版本之一，YOLO v5 相比于原始的 YOLO 算法，在检测准确性和速度上都有显著的提升，使得它成为一个广泛使用的目标检测算法。然后，YOLO 算法继续进行改进和优化，发展出了后续版本的 YOLO，如 YOLO v3、YOLO v4 和 YOLO v5，以进一步提升目标检测的性能。这些改进版本在网络结构、特征提取、预测机制等方面进行了创新和改进，以满足不同应用场景下的需求。

在 Python 中使用 YOLO v5 前需要先通过如下命令安装：

```
pip install ultralytics
```

作为初学者来说，可以直接使用 YOLO v5 提供的预训练模型。预训练的 YOLO v5 模型可以从多个来源下载，以下是一些常用的下载来源：

- Darknet 官方网站：YOLO v5 是由 Joseph Redmon 开发的 Darknet 框架的一部分。可以通过 Darknet 官方网站下载 YOLO v5 的预训练模型。访问链接：https://pjreddie.com/darknet/yolo/。
- YOLO 官方 GitHub 页面：YOLO 官方 GitHub 仓库也提供了 YOLO v5 的预训练模型。可以在 https://github.com/pjreddie/darknet/releases 页面找到预训练模型的下载链接。
- 第三方资源：除了官方来源，还有一些第三方资源库和社区提供了 YOLO v5 的预训练模型的下载。一些常见的资源库包括 Model Zoo、PyTorch Hub、Hugging Face Model Hub 等。

在下载预训练的 YOLO v5 模型时，请确保使用可信赖的来源，并查看模型的许可和使用条款。预训练模型的下载可能需要注册、登录或同意特定的使用条件，具体取决于下载来源。

注意：由于 YOLO v5 模型在不同的深度学习框架中可能具有不同的实现和权重格式，因此需要确保下载的模型与使用的深度学习框架兼容。一般来说，官方提供的预训练模型会与官方支持的框架兼容性较好。

6.2.2　基于 YOLO v5 的训练、验证和预测

看下面的例子，功能是使用 YOLO v5 实现训练、验证和预测功能，让大家初步了解 YOLO v5 实现图像目标检测的基本功能。

实例 6-1：使用 YOLO v5 实现模型训练、验证和预测

源码路径： daima\6\yolov5-master

1. 目标检测

目标检测是一种计算机视觉任务，旨在从图像或视频中定位和识别出特定对象的位置。通常使用深度学习模型来进行目标检测，例如使用卷积神经网络或相关的模型，如 YOLO（you only look once）和 Faster R-CNN（region-based convolutional neural networks）。在本实例中，编写文件 detect.py 实现目标检测任务，这是一个运行 YOLO v5 推理的脚本，可以在各种来源上

进行推理，自动从最新的 YOLO v5 发布中下载模型，并将结果保存到"runs/detect"目录下。文件 detect.py 的具体实现流程如下：

（1）定义了一个名为 run 的函数，该函数用于运行目标检测任务。具体实现代码如下：

```
def run(
        weights=ROOT / 'yolov5s.pt',
        source=ROOT / 'data/images',
        data=ROOT / 'data/coco128.yaml',
        imgsz=(640, 640),
        conf_thres=0.25,
        iou_thres=0.45,
        max_det=1000,
        device='',
        view_img=False,
        save_txt=False,
        save_conf=False,
        save_crop=False,
        nosave=False,
        classes=None,
        agnostic_nms=False,
        augment=False,
        visualize=False,
        update=False,
        project=ROOT / 'runs/detect',
        name='exp',
        exist_ok=False,
        line_thickness=3,
        hide_labels=False,
        hide_conf=False,
        half=False,
        dnn=False,
        vid_stride=1,
):
```

在上述代码中，各个参数的具体说明如下：

- weights：模型的路径或 Triton URL，默认值为 yolov5s.pt，表示模型的权重文件路径。
- source：推理的来源，可以是文件、目录、URL、通配符、屏幕截图或者摄像头。默认值为 data/images，表示推理的来源为 data/images 目录。
- data：数据集的配置文件路径，默认值为 data/coco128.yaml，表示使用 COCO128 数据集的配置文件。
- imgsz：推理时的图像尺寸，默认为(640, 640)，表示推理时将图像调整为高度和宽度都为 640 的尺寸。
- conf_thres：置信度阈值，默认值为 0.25，表示只保留置信度大于该阈值的检测结果。
- iou_thres：NMS（非极大值抑制）的 IOU（交并比）阈值，默认值为 0.45，用于去除重叠度较高的重复检测结果。
- max_det：每张图像的最大检测数量，默认值为 1 000，表示每张图像最多保留 1 000 个检测结果。
- device：设备类型，默认为空字符串，表示使用默认设备（GPU 或 CPU）进行推理。
- view_img：是否显示结果图像，默认值为 False，表示不显示结果图像。
- save_txt：是否将结果保存为文本文件，默认值为 False。

- save_conf：是否将置信度保存在保存的文本标签中，默认值为 False。
- save_crop：是否保存裁剪的预测框，默认值为 False。
- nosave：是否禁止保存图像或视频，默认值为 False。
- classes：根据类别进行过滤，默认为 None，表示不进行类别过滤。
- agnostic_nms：是否使用类别不可知的 NMS，默认值为 False。
- augment：是否进行增强推理，默认值为 False。
- visualize：是否可视化特征，默认值为 False。
- update：是否更新所有模型，默认值为 False。
- project：保存结果的项目路径，默认为 runs/detect。
- name：保存结果的名称，默认为 exp。
- exist_ok：是否允许存在的项目/名称，如果为 True，则不递增项目/名称，默认值为 False。
- line_thickness：边界框线条的粗细，默认为 3 个像素。
- hide_labels：是否隐藏标签，默认值为 False。
- hide_conf：是否隐藏置信度，默认值为 False。
- half：是否使用 FP16 的半精度推理，默认值为 False。
- dnn：是否使用 OpenCV DNN 进行 ONNX 推理，默认值为 False。
- vid_stride：视频帧率步长，默认值为 1。
- 这些参数可以根据需要进行调整，以适应不同的目标检测场景和要求。

（2）运行目标检测推理，并根据不同的输入来源进行相应的处理和操作。具体实现代码如下：

```
        source = str(source)
        save_img = not nosave and not source.endswith('.txt')  # 是否保存推理图像
        is_file = Path(source).suffix[1:] in (IMG_FORMATS + VID_FORMATS)  # 是否为文件路径
        is_url = source.lower().startswith(('rtsp://', 'rtmp://', 'http://', 'https://'))
# 是否为 URL 地址
        webcam = source.isnumeric() or source.endswith('.streams') or (is_url and not
is_file)  # 是否为摄像头或流
        screenshot = source.lower().startswith('screen')  # 是否为屏幕截图
        if is_url and is_file:
            source = check_file(source)  # 下载文件

        save_dir = increment_path(Path(project) / name, exist_ok=exist_ok)  # 创建保存目录
        (save_dir / 'labels' if save_txt else save_dir).mkdir(parents=True, exist_ok=True)
# 创建标签目录（如果需要保存标签）

        # 加载模型
        device = select_device(device)  # 选择设备
        model = DetectMultiBackend(weights, device=device, dnn=dnn, data=data, fp16=half)
# 加载模型
        stride, names, pt = model.stride, model.names, model.pt  # 获取模型的步幅、类别名
称、网络架构
        imgsz = check_img_size(imgsz, s=stride)  # 检查图像大小

        # 数据加载器
        bs = 1  # 批大小
        if webcam:
            view_img = check_imshow(warn=True)  # 检查是否显示图像
            dataset = LoadStreams(source, img_size=imgsz, stride=stride, auto=pt, vid_stride=
vid_stride)  # 加载视频流
```

```python
            bs = len(dataset)  # 设置批大小为视频流的数量
        elif screenshot:
            dataset = LoadScreenshots(source, img_size=imgsz, stride=stride, auto=pt)  # 加载屏幕截图
        else:
            dataset = LoadImages(source, img_size=imgsz, stride=stride, auto=pt, vid_stride=vid_stride)  # 加载图像
        vid_path, vid_writer = [None] * bs, [None] * bs  # 初始化视频路径和视频写入器

        # 运行推理
        model.warmup(imgsz=(1 if pt or model.triton else bs, 3, *imgsz))  # 模型热身
        seen, windows, dt = 0, [], (Profile(), Profile(), Profile())  # 初始化变量
        for path, im, im0s, vid_cap, s in dataset:
            with dt[0]:
                im = torch.from_numpy(im).to(model.device)  # 转换图像到张量
                im = im.half() if model.fp16 else im.float()  # 转换张量数据类型
                im /= 255  # 数据归一化
                if len(im.shape) == 3:
                    im = im[None]  # 扩展为批大小维度

            # 推理
            with dt[1]:
                visualize = increment_path(save_dir / Path(path).stem, mkdir=True) if visualize else False  # 是否可视化
                pred = model(im, augment=augment, visualize=visualize)  # 进行推理

            # 非极大值抑制
            with dt[2]:
                pred = non_max_suppression(pred, conf_thres, iou_thres, classes, agnostic_nms, max_det=max)  # 非极大值抑制
```

对上述代码的具体说明如下：

- 首先，对输入的 source 进行了一系列判断和处理。将 source 转换为字符串类型，并根据条件判断是否保存推理图像。接着，判断 source 是文件路径还是 URL，以及是否为摄像头或屏幕截图。如果 source 是 URL 且为文件路径，则会进行文件下载操作。
- 创建保存结果的目录，根据 project 和 name 参数生成保存结果的路径，并在指定路径下创建目录。如果 save_txt 为 True，则在目录下创建一个名为'labels'的子目录，否则直接创建主目录。
- 加载模型并选择设备。根据指定的设备类型，选择相应的设备进行推理。同时，实例化 DetectMultiBackend 类的对象 model，并获取其步长（stride）、类别名称（names）和模型（pt）。
- 检查图像尺寸并创建数据加载器。根据不同的输入来源，选择相应的数据加载器对象：LoadStreams 用于摄像头输入；LoadScreenshots 用于屏幕截图输入；LoadImages 用于图像或视频输入。同时，根据是否为摄像头输入，确定批处理大小（bs）。
- 进行推理过程。首先，通过调用 model.warmup()方法进行模型预热，其中输入图像尺寸根据模型类型进行调整。然后，使用迭代器遍历数据加载器，获取输入图像及相关信息。将图像转换为 PyTorch 张量，并根据模型的精度要求进行数据类型和范围的调整。如果输入图像维度为三维，则添加一个批处理维度。
- 进行推理过程。根据是否需要可视化结果，确定是否保存推理结果的路径，并调用 model

对象的__call__()方法进行推理。得到推理结果后，根据设定的置信度阈值、IOU阈值和其他参数，进行非最大值抑制（NMS）操作。
- 最后，对推理结果进行处理。包括计算推理的时间、保存推理结果图像和输出结果信息。其中，推理时间分为三个阶段，即数据准备、模型推理和NMS操作。保存推理结果图像的路径根据是否需要可视化和输入图像的文件名进行生成。根据设置的参数决定是否将推理结果保存为文本文件。

2．训练

在深度学习中，"train"表示训练模型的过程。训练是指通过给定的输入数据和相应的标签，使用梯度下降等优化算法来调整模型的参数，使其逐渐适应给定的任务。训练过程通常包括前向传播、计算损失函数、反向传播和参数更新等步骤。通过反复迭代训练数据集，模型可以逐渐学习到数据中的模式和特征，从而提高在给定任务上的性能。在本实例中，编写文件train.py实现训练功能。具体实现流程如下：

（1）编写函数train()用于训练模型。首先通过下面的代码进行训练前的准备工作，包括解析参数、创建目录、加载超参数、保存运行设置和创建日志记录器等。它为后续的训练过程提供了必要的信息和设置。

```
def train(hyp, opt, device, callbacks):
    # 解析参数
    save_dir = Path(opt.save_dir)
    epochs = opt.epochs
    batch_size = opt.batch_size
    weights = opt.weights
    single_cls = opt.single_cls
    evolve = opt.evolve
    data = opt.data
    cfg = opt.cfg
    resume = opt.resume
    noval = opt.noval
    nosave = opt.nosave
    workers = opt.workers
    freeze = opt.freeze

    # 执行回调函数的预训练例行程序
    callbacks.run('on_pretrain_routine_start')

    # 创建目录
    w = save_dir / 'weights'  # 权重目录
    (w.parent if evolve else w).mkdir(parents=True, exist_ok=True)  # 创建目录
    last, best = w / 'last.pt', w / 'best.pt'

    # 超参数
    if isinstance(hyp, str):
        with open(hyp, errors='ignore') as f:
            hyp = yaml.safe_load(f)  # 加载超参数字典
    LOGGER.info(colorstr('hyperparameters: ') + ', '.join(f'{k}={v}' for k, v in hyp.items()))
    opt.hyp = hyp.copy()  # 保存超参数到检查点

    # 保存运行设置
    if not evolve:
```

```
            yaml_save(save_dir / 'hyp.yaml', hyp)
            yaml_save(save_dir / 'opt.yaml', vars(opt))

    # 日志记录器
    data_dict = None
    if RANK in {-1, 0}:
        loggers = Loggers(save_dir, weights, opt, hyp, LOGGER)  # 日志记录器实例

        # 注册动作
        for k in methods(loggers):
            callbacks.register_action(k, callback=getattr(loggers, k))

        # 处理自定义数据集的工件链接
        data_dict = loggers.remote_dataset
        if resume:  # 从远程工件恢复运行
            weights, epochs, hyp, batch_size = opt.weights, opt.epochs, opt.hyp, opt.batch_size
```

对上述代码的具体说明如下：
- 将函数的参数和选项进行解析和赋值。包括 hyp（超参数路径或字典）、opt（选项参数对象）、device（设备）、callbacks（回调函数集合）等。
- 创建保存目录，并设置权重目录 w。
- 加载超参数。如果 hyp 是一个字符串，则从文件中加载超参数字典。然后将超参数保存在 opt.hyp 中，以便在训练过程中保存到检查点。
- 如果不是进化训练（evolve=False），则保存运行设置。将超参数保存为 hyp.yaml 文件，将选项参数保存为 opt.yaml 文件。
- 创建日志记录器（Loggers）实例。如果当前进程是主进程（RANK=-1 或 RANK=0），则创建日志记录器并注册回调函数。
- 如果是从远程 artifact 恢复运行，则更新权重、轮数、超参数和批大小。

（2）继续训练模型前的准备工作，包括配置设置、加载模型、冻结层、图像尺寸、批大小、优化器、学习率调度器、EMA 指数滑动平均、恢复训练等。具体实现代码如下：

```
def train(hyp, opt, device, callbacks):
    # 解析参数
    save_dir = Path(opt.save_dir)
    epochs = opt.epochs
    batch_size = opt.batch_size
    weights = opt.weights
    single_cls = opt.single_cls
    evolve = opt.evolve
    data = opt.data
    cfg = opt.cfg
    resume = opt.resume
    noval = opt.noval
    nosave = opt.nosave
    workers = opt.workers
    freeze = opt.freeze

    # 执行回调函数的预训练例行程序
    callbacks.run('on_pretrain_routine_start')

    # 创建目录
```

```python
w = save_dir / 'weights'  # 权重目录
(w.parent if evolve else w).mkdir(parents=True, exist_ok=True)  # 创建目录
last, best = w / 'last.pt', w / 'best.pt'

# 超参数
if isinstance(hyp, str):
    with open(hyp, errors='ignore') as f:
        hyp = yaml.safe_load(f)  # 加载超参数字典
LOGGER.info(colorstr('hyperparameters: ') + ', '.join(f'{k}={v}' for k, v in hyp.items()))
opt.hyp = hyp.copy()  # 保存超参数到检查点

# 保存运行设置
if not evolve:
    yaml_save(save_dir / 'hyp.yaml', hyp)
    yaml_save(save_dir / 'opt.yaml', vars(opt))

# 日志记录器
data_dict = None
if RANK in {-1, 0}:
    loggers = Loggers(save_dir, weights, opt, hyp, LOGGER)  # 日志记录器实例

    # 注册动作
    for k in methods(loggers):
        callbacks.register_action(k, callback=getattr(loggers, k))

    # 处理自定义数据集的工件链接
    data_dict = loggers.remote_dataset
    if resume:  # 从远程工件恢复运行
        weights, epochs, hyp, batch_size = opt.weights, opt.epochs, opt.hyp, opt.batch_size
```

（3）使用深度学习框架 PyTorch 训练目标检测模型，具体实现代码如下：

```python
train_loader, dataset = create_dataloader(train_path,
                                          imgsz,
                                          batch_size // WORLD_SIZE,
                                          gs,
                                          single_cls,
                                          hyp=hyp,
                                          augment=True,
                                          cache=None if opt.cache == 'val' else opt.cache,
                                          rect=opt.rect,
                                          rank=LOCAL_RANK,
                                          workers=workers,
                                          image_weights=opt.image_weights,
                                          quad=opt.quad,
                                          prefix=colorstr('train: '),
                                          shuffle=True,
                                          seed=opt.seed)
labels = np.concatenate(dataset.labels, 0)
mlc = int(labels[:, 0].max())  # max label class
assert mlc < nc, f'Label class {mlc} exceeds nc={nc} in {data}. Possible class labels are 0-{nc - 1}'

# 处理进程 0
if RANK in {-1, 0}:
    val_loader = create_dataloader(val_path,
```

```
                                        imgsz,
                                        batch_size // WORLD_SIZE * 2,
                                        gs,
                                        single_cls,
                                        hyp=hyp,
                                        cache=None if noval else opt.cache,
                                        rect=True,
                                        rank=-1,
                                        workers=workers * 2,
                                        pad=0.5,
                                        prefix=colorstr('val: '))[0]

        if not resume:
            if not opt.noautoanchor:
                check_anchors(dataset, model=model, thr=hyp['anchor_t'], imgsz=imgsz)   # run AutoAnchor
            model.half().float()   # pre-reduce anchor precision

        callbacks.run('on_pretrain_routine_end', labels, names)

    # DDP 模式
    if cuda and RANK != -1:
        model = smart_DDP(model)   # 智能分布式数据并行模式

    # 模型属性
    nl = de_parallel(model).model[-1].nl   # 检测层数量（用于调整超参数）
    hyp['box'] *= 3 / nl   # 缩放至层
    hyp['cls'] *= nc / 80 * 3 / nl   # 缩放至类别和层
    hyp['obj'] *= (imgsz / 640) ** 2 * 3 / nl   # 缩放至图像尺寸和层
    hyp['label_smoothing'] = opt.label_smoothing   # 标签平滑参数
    model.nc = nc   # 将类别数附加到模型
    model.hyp = hyp   # 将超参数附加到模型
    model.class_weights = labels_to_class_weights(dataset.labels, nc).to(device) * nc   # 将类别权重附加到模型
    model.names = names   # 将类别名称附加到模型

    # 开始训练
    t0 = time.time()   # 记录开始时间
    nb = len(train_loader)   # 批次数量
    nw = max(round(hyp['warmup_epochs'] * nb), 100)   # 热身迭代次数，最大值为（3个周期，100次迭代）
    last_opt_step = -1   # 上次优化步骤
    maps = np.zeros(nc)   # 每个类别的 mAP
    results = (0, 0, 0, 0, 0, 0, 0)   # 结果（P, R, mAP@.5, mAP@.5-.95, val_loss(box, obj, cls))
    scheduler.last_epoch = start_epoch - 1   # 不移动
    scaler = torch.cuda.amp.GradScaler(enabled=amp)   # 梯度缩放器
    stopper, stop = EarlyStopping(patience=opt.patience), False   # 早停参数
    compute_loss = ComputeLoss(model)   # 初始化损失类
    callbacks.run('on_train_start')   # 运行训练开始回调
    LOGGER.info(f'图像尺寸 {imgsz} 训练，{imgsz} 验证\n'
                f'使用 {train_loader.num_workers * WORLD_SIZE} 数据加载器工作进程\n'
                f"记录结果到 {colorstr('bold', save_dir)}\n"
                f'开始训练 {epochs} 个周期...')
    for epoch in range(start_epoch, epochs):   # 每个周期 ------------------------------
        callbacks.run('on_train_epoch_start')   # 运行每个周期开始回调
```

```python
        model.train()  # 模型训练模式

        # 更新图像权重（可选，仅单GPU）
        if opt.image_weights:
            cw = model.class_weights.cpu().numpy() * (1 - maps) ** 2 / nc  # 类别权重
            iw = labels_to_image_weights(dataset.labels, nc=nc, class_weights=cw)  # 图像权重
            dataset.indices = random.choices(range(dataset.n), weights=iw, k=dataset.n)  # 随机加权索引

        # 更新马赛克边界（可选）
        # b = int(random.uniform(0.25 * imgsz, 0.75 * imgsz + gs) // gs * gs)
        # dataset.mosaic_border = [b - imgsz, -b]  # 高度、宽度边界

        mloss = torch.zeros(3, device=device)  # 平均损失
        if RANK != -1:
            train_loader.sampler.set_epoch(epoch)  # 设置epoch
        pbar = enumerate(train_loader)  # 创建进度条
        LOGGER.info(('\n' + '%11s' * 7) % ('周期', 'GPU 内存', '框损失', '目标损失', '类别损失', '实例', '尺寸'))
        if RANK in {-1, 0}:
            pbar = tqdm(pbar, total=nb, bar_format=TQDM_BAR_FORMAT)  # 进度条
        optimizer.zero_grad()
        for i, (imgs, targets, paths, _) in pbar:  # 批次 -------------------------------
            callbacks.run('on_train_batch_start')  # 运行批次开始回调
            ni = i + nb * epoch  # 整合批次数（自训练开始以来）
            imgs = imgs.to(device, non_blocking=True).float

            # 热身
            if ni <= nw:
                xi = [0, nw]  # x 插值
                # compute_loss.gr = np.interp(ni, xi, [0.0, 1.0])  # iou损失比率(obj_loss = 1.0 或 iou)
                accumulate = max(1, np.interp(ni, xi, [1, nbs / batch_size]).round())
                for j, x in enumerate(optimizer.param_groups):
                    # 偏置学习率从0.1下降到lr0，所有其他学习率从0.0上升到lr0
                    x['lr'] = np.interp(ni, xi, [hyp['warmup_bias_lr'] if j == 0 else 0.0, x['initial_lr'] * lf(epoch)])
                    if 'momentum' in x:
                        x['momentum'] = np.interp(ni, xi, [hyp['warmup_momentum'], hyp['momentum']])

            # 多尺度
            if opt.multi_scale:
                sz = random.randrange(int(imgsz * 0.5), int(imgsz * 1.5) + gs) // gs * gs  # 大小
                sf = sz / max(imgs.shape[2:])  # 缩放因子
                if sf != 1:
                    ns = [math.ceil(x * sf / gs) * gs for x in imgs.shape[2:]]  # 新形状（拉伸至gs倍数）
                    imgs = nn.functional.interpolate(imgs, size=ns, mode='bilinear', align_corners=False)

            # 前向传播
            with torch.cuda.amp.autocast(amp):
                pred = model(imgs)  # 前向传播
```

```
                    loss, loss_items = compute_loss(pred, targets.to(device))  # 损失按
batch_size缩放
                    if RANK != -1:
                        loss *= WORLD_SIZE  # DDP 模式下在设备之间平均梯度
                    if opt.quad:
                        loss *= 4.

                    # 反向传播
                    scaler.scale(loss).backward()

                    # 优化 - https://pytorch.org/docs/master/notes/amp_examples.html
                    if ni - last_opt_step >= accumulate:
                        scaler.unscale_(optimizer)  # 取消缩放梯度
                        torch.nn.utils.clip_grad_norm_(model.parameters(), max_norm=6.0)
# 限制梯度
                        scaler.step(optimizer)  # 优化器步骤
                        scaler.update()
                        optimizer.zero_grad()
                        if ema:
                            ema.update(model)
                        last_opt_step = ni

                    # 日志
                    if RANK in {-1, 0}:
                        mloss = (mloss * i + loss_items) / (i + 1)  # 更新平均损失
                        mem = f'{torch.cuda.memory_reserved() / 1E9 if torch.cuda.is_available()
else 0:.3g}G'  # (GB)
                        pbar.set_description(('%11s' * 2 + '%11.4g' * 5) %
                                             (f'{epoch}/{epochs - 1}', mem, *mloss, targets.
shape[0], imgs.shape[-1]))
                        callbacks.run('on_train_batch_end', model, ni, imgs, targets, paths,
list(mloss))
                        if callbacks.stop_training:
                            return
                    # 结束批次 ------------------------------
```

对上述代码的具体说明如下：
- 创建训练数据集的数据加载器，并获取数据集中的标签信息。
- 根据训练模式进行不同的操作。如果是分布式训练模式且使用了多个 GPU，则使用 torch.nn.DataParallel 将模型包装起来，以实现多 GPU 并行训练。如果开启了同步 BatchNorm，并且不是分布式训练模式，则将模型中的 BatchNorm 层转换为 SyncBatchNorm，以实现跨 GPU 同步。
- 对模型的一些属性进行设置，包括调整损失函数的权重，附加类别权重、类别名称等信息到模型上。
- 开始训练过程，包括设置一些参数（如学习率调整、early stopping 等），创建优化器、损失函数等。进行多个 epoch 的训练，每个 epoch 中遍历数据集的批次进行训练。在训练过程中，根据设置的参数进行一些预处理操作，如数据增强、图片尺寸调整等。将数据输入模型进行前向传播，得到预测结果，并计算损失函数。
- 进行反向传播和优化器更新操作。如果达到一定的条件（如累计一定数量的批次），则进行一次优化器的更新操作。

在训练过程中，会输出一些训练信息，如当前的 epoch、GPU 内存占用、损失值等。同时，会调用一些回调函数，如在每个批次开始前和结束后运行的函数。整个训练过程会持续进行多个 epoch，直到达到指定的训练轮数为止。

（4）训练模型，使用循环训练目标检测模型。具体实现代码如下：

```
            fi = fitness(np.array(results).reshape(1, -1))  # [P, R, mAP@.5, mAP@.5-.95]
的加权组合
            stop = stopper(epoch=epoch, fitness=fi)  # 检查是否早停
            if fi > best_fitness:
                best_fitness = fi
            log_vals = list(mloss) + list(results) + lr
            callbacks.run('on_fit_epoch_end', log_vals, epoch, best_fitness, fi)

            # 保存模型
            if (not nosave) or (final_epoch and not evolve):  # 如果需要保存
                ckpt = {
                    'epoch': epoch,
                    'best_fitness': best_fitness,
                    'model': deepcopy(de_parallel(model)).half(),  # 深拷贝并转为半精度
                    'ema': deepcopy(ema.ema).half(),  # 深拷贝并转为半精度
                    'updates': ema.updates,
                    'optimizer': optimizer.state_dict(),
                    'opt': vars(opt),
                    'git': GIT_INFO,  # 如果是 git repo，则包含{remote, branch, commit}
                    'date': datetime.now().isoformat()}

                # 保存最新的、最好的，并删除中间保存的模型
                torch.save(ckpt, last)
                if best_fitness == fi:
                    torch.save(ckpt, best)
                if opt.save_period > 0 and epoch % opt.save_period == 0:
                    torch.save(ckpt, w / f'epoch{epoch}.pt')
                del ckpt
                callbacks.run('on_model_save', last, epoch, final_epoch, best_fitness, fi)

        # 早停
        if RANK != -1:  # 如果是 DDP 训练
            broadcast_list = [stop if RANK == 0 else None]
            dist.broadcast_object_list(broadcast_list, 0)  # 将'stop'广播给所有 rank
            if RANK != 0:
                stop = broadcast_list[0]
        if stop:
            break  # 必须中断所有 DDP ranks

    # 结束 epoch 循环
    # 结束训练
    if RANK in {-1, 0}:
        LOGGER.info(f'\n{epoch - start_epoch + 1} epochs completed in {(time.time() - t0) / 3600:.3f} hours.')
        for f in last, best:
            if f.exists():
                strip_optimizer(f)  # 去除优化器
                if f is best:
                    LOGGER.info(f'\nValidating {f}...')
                    results, _, _ = validate.run(
```

```
                    data_dict,
                    batch_size=batch_size // WORLD_SIZE * 2,
                    imgsz=imgsz,
                    model=attempt_load(f, device).half(),
                    iou_thres=0.65 if is_coco else 0.60,  # pycocotools 的最佳 iou 为 0.65
                    single_cls=single_cls,
                    dataloader=val_loader,
                    save_dir=save_dir,
                    save_json=is_coco,
                    verbose=True,
                    plots=plots,
                    callbacks=callbacks,
                    compute_loss=compute_loss)  # 用验证集验证最佳模型并绘制图表
                if is_coco:
                    callbacks.run('on_fit_epoch_end', list(mloss) + list(results) + lr, epoch, best_fitness, fi)

        callbacks.run('on_train_end', last, best, epoch, results)

    torch.cuda.empty_cache()  # 清空 CUDA 缓存
    return results
```

对上述代码的具体说明如下：

- fi = fitness(np.array(results).reshape(1, −1))：计算模型在验证集上的综合指标。results 是一个包含模型在验证集上表现的元组，通过调用 fitness 函数计算综合指标。

- stop = stopper(epoch=epoch, fitness=fi)：判断是否满足停止训练的条件。stopper 是一个用于判断是否进行 early stopping 的对象，根据当前的训练轮数 epoch 和综合指标 fitness 来判断是否停止训练。

- if fi > best_fitness: best_fitness = fi：更新最佳的综合指标 best_fitness，如果当前的综合指标大于最佳指标。

- log_vals = list(mloss) + list(results) + lr：将当前的损失值、验证集上的结果和学习率组成一个列表 log_vals，用于记录训练过程中的日志。

- ckpt = {...}：创建一个字典 ckpt，保存训练过程中的相关信息，包括当前的轮数 epoch、最佳综合指标 best_fitness、模型参数、优化器状态等。

- torch.save(ckpt, last)：将 ckpt 保存到文件 last，这里保存的是最后一轮的模型。

- if best_fitness == fi: torch.save(ckpt, best)：如果当前的综合指标等于最佳指标，将 ckpt 保存到文件 best，这里保存的是最佳的模型。

- if opt.save_period > 0 and epoch % opt.save_period == 0: torch.save(ckpt, w / f'epoch{epoch}.pt')：如果设置了保存周期 save_period 且当前轮数是保存周期的倍数，将 ckpt 保存到文件 epoch{epoch}.pt，用于定期保存模型。

- callbacks.run('on_model_save', last, epoch, final_epoch, best_fitness, fi)：运行回调函数 on_model_save，将保存的模型文件路径、当前轮数、是否是最后一轮、最佳综合指标和当前综合指标等参数传递给回调函数。

- if stop: break：如果满足停止训练的条件，跳出训练循环，结束训练。

- if RANK in {−1, 0}: LOGGER.info(f'\n{epoch − start_epoch + 1} epochs completed in {(time.time() − t0) / 3600:.3f} hours.')：如果是主进程（RANK 为-1 或 0），打印训练完成

的信息，包括训练轮数和所花费的时间。
- for f in last, best: ...：遍历最后一轮的模型和最佳模型。
- if f.exists(): ...：如果模型文件存在。
- strip_optimizer(f)：去除模型文件中的优化器信息。
- if f is best: ...：如果是最佳模型，运行验证函数对模型进行评估。
- callbacks.run('on_fit_epoch_end', list(mloss) + list(results) + lr, epoch, best_fitness, fi)：运行回调函数 on_fit_epoch_end，将损失值、验证集结果和学习率等参数传递给回调函数。
- callbacks.run('on_train_end', last, best, epoch, results)：运行回调函数 on_train_end，将最后一轮模型、最佳模型、当前轮数和验证集结果等参数传递给回调函数。
- torch.cuda.empty_cache()：清空 GPU 缓存。

3. val 模型验证

在深度学习中，"val" 通常指的是验证数据集。验证数据集是在训练模型过程中用于评估模型性能和调整超参数的数据集。训练过程通常分为训练集、验证集和测试集三部分。验证集用于在训练过程中评估模型的性能，并根据验证结果调整模型的超参数，以优化模型的泛化能力。与训练集用于训练模型不同，验证集的目的是评估模型在未见过的数据上的性能，以避免过拟合和选择合适的模型。在本项目中，编写文件 val.py 实现模型验证功能，具体实现流程如下：

（1）编写函数 save_one_txt()，功能是将目标检测模型的预测结果保存到文本文件中。具体实现流程如下：通过计算归一化增益，将预测框的坐标从 xyxy 格式转换为归一化的 xywh 格式。先遍历每个预测框，并根据是否保存置信度确定输出格式再将结果写入文本文件中，每个预测结果占一行。

函数 save_one_txt() 的具体实现代码如下：

```
def save_one_txt(predn, save_conf, shape, file):
    # 保存单个文本结果
    gn = torch.tensor(shape)[[1, 0, 1, 0]]  # 归一化增益 whwh
    for *xyxy, conf, cls in predn.tolist():
        xywh = (xyxy2xywh(torch.tensor(xyxy).view(1, 4)) / gn).view(-1).tolist()  # 归一化的 xywh
        line = (cls, *xywh, conf) if save_conf else (cls, *xywh)  # 标签格式
        with open(file, 'a') as f:
            f.write(('%g ' * len(line)).rstrip() % line + '\n')
```

（2）编写函数 save_one_json() 的功能是将目标检测模型的预测结果保存为 JSON 格式的文件，具体实现流程如下：
① 根据输入文件路径提取图像 ID。
② 将预测框的坐标从 xyxy 格式转换为 xywh 格式，并将中心坐标转换为左上角坐标。
③ 遍历每个预测框，将预测结果以字典的形式添加到列表中。
④ 字典包括图像 ID、类别 ID、边界框坐标和置信度。

函数 save_one_json() 的具体实现代码如下：

```
def save_one_json(predn, jdict, path, class_map):
    # 保存单个 JSON 结果 {"image_id": 42, "category_id": 18, "bbox": [258.15, 41.29, 348.26, 243.78], "score": 0.236}
    image_id = int(path.stem) if path.stem.isnumeric() else path.stem
    box = xyxy2xywh(predn[:, :4])  # 转换为 xywh 格式
    box[:, :2] -= box[:, 2:] / 2  # 将 xy 中心转换为左上角
```

```
    for p, b in zip(predn.tolist(), box.tolist()):
        jdict.append({
            'image_id': image_id,
            'category_id': class_map[int(p[5])],
            'bbox': [round(x, 3) for x in b],
            'score': round(p[4], 5)})
```

（3）编写函数 process_batch()，功能是根据预测框和标签框计算正确的预测矩阵，具体实现流程如下：

① 创建一个全零矩阵 correct，用于存储预测框和标签框之间的匹配情况。

② 计算预测框和标签框之间的 IoU（交并比）。

③ 对于每个 IoU 阈值，筛选出 IoU 大于阈值且类别匹配的预测框。

④ 将匹配的结果记录在 correct 矩阵中，用布尔值表示匹配情况。

⑤ 返回一个 correct 张量，其中每一行表示一个预测框，每一列表示一个 IoU 阈值，值为 True 表示匹配正确。

函数 process_batch()的具体实现代码如下：

```
def process_batch(detections, labels, iouv):
    correct = np.zeros((detections.shape[0], iouv.shape[0])).astype(bool)
    iou = box_iou(labels[:, 1:], detections[:, :4])
    correct_class = labels[:, 0:1] == detections[:, 5]
    for i in range(len(iouv)):
        x = torch.where((iou >= iouv[i]) & correct_class)  # IoU > 阈值且类别匹配
        if x[0].shape[0]:
            matches = torch.cat((torch.stack(x, 1), iou[x[0], x[1]][:, None]), 1).cpu().numpy()  # [label, detect, iou]
            if x[0].shape[0] > 1:
                matches = matches[matches[:, 2].argsort()[::-1]]
                matches = matches[np.unique(matches[:, 1], return_index=True)[1]]
                # matches = matches[matches[:, 2].argsort()[::-1]]
                matches = matches[np.unique(matches[:, 0], return_index=True)[1]]
            correct[matches[:, 1].astype(int), i] = True
    return torch.tensor(correct, dtype=torch.bool, device=iouv.device)
```

通过运行下面的命令可以展示识别结果：

```
python detect.py --weights yolov5s.pt --img 640 --conf 0.25 --source data/images
```

6.3 语义分割

语义分割（semantic segmentation）是计算机视觉领域中的一项任务，旨在将图像中的每个像素进行分类，将图像分割成具有语义信息的不同区域。与目标检测只关注检测和定位图像中的物体不同，语义分割不仅要检测物体，还要对每个像素进行分类，将每个像素标记为属于不同的语义类别。

6.3.1 什么是语义分割

语义分割的目标是为图像中的每个像素分配一个语义标签，通常使用颜色或类别编号来表示不同的语义类别。例如，在一张街景图像中，语义分割可以将图像分割成道路、汽车、行人、建筑物等不同的语义区域。

语义分割对于许多计算机视觉任务非常重要，例如自动驾驶中的场景理解、医学图像分析中的病变检测、智能视频监控中的目标跟踪等。通过准确地理解图像中的语义信息，可以为这些任务提供更丰富的场景理解和更精准的分析结果。

为了实现语义分割，常用的方法是使用深度学习模型，特别是卷积神经网络。这些模型能够学习到图像的特征表示，并通过卷积和上采样等操作实现对每个像素的分类。近年来，基于深度学习的语义分割方法取得了显著的进展，并在许多任务和应用中取得了优秀的性能。

例如，下面是一个使用卷积神经网络进行图像分割的例子，在 Python 程序中使用深度学习框架 PyTorch 实现图像分割功能。

实例6-2：使用卷积神经网络进行图像分割

源码路径： daima\6\juanfen.py

```python
import torch
import torch.nn as nn

# 定义卷积神经网络模型
class SegmentationModel(nn.Module):
    def __init__(self, num_classes):
        super(SegmentationModel, self).__init__()
        # 定义网络的层和操作
        self.conv1 = nn.Conv2d(3, 64, kernel_size=3, stride=1, padding=1)
        self.relu = nn.ReLU()
        self.conv2 = nn.Conv2d(64, 64, kernel_size=3, stride=1, padding=1)
        self.conv3 = nn.Conv2d(64, num_classes, kernel_size=1, stride=1)

    def forward(self, x):
        x = self.relu(self.conv1(x))
        x = self.relu(self.conv2(x))
        x = self.conv3(x)
        return x

# 创建模型实例
num_classes = 2  # 两个类别：前景和背景
model = SegmentationModel(num_classes)

# 加载图像数据和标签
input_image = torch.randn(1, 3, 256, 256)  # 输入图像，假设大小为256×256，通道数为3
target_mask = torch.randint(0, num_classes, (1, 256, 256))  # 目标分割掩码，假设大小为256×256

# 定义损失函数和优化器
criterion = nn.CrossEntropyLoss()
optimizer = torch.optim.SGD(model.parameters(), lr=0.001, momentum=0.9)

# 训练模型
num_epochs = 10
for epoch in range(num_epochs):
    # 前向传播
    output = model(input_image)

    # 计算损失
    loss = criterion(output, target_mask)
```

```
# 反向传播和优化
optimizer.zero_grad()
loss.backward()
optimizer.step()

# 输出当前训练状态
print(f'Epoch [{epoch+1}/{num_epochs}], Loss: {loss.item()}')
```

在上述代码中定义了一个简单的卷积神经网络模型，包含几个卷积层和激活函数。模型的输出是一个与输入图像大小相同的张量，每个像素点都表示对应的类别。在训练过程中，使用交叉熵损失函数（CrossEntropyLoss）来计算模型输出与目标分割掩码之间的差异，并使用随机梯度下降（SGD）优化器更新模型的参数。执行后会输出：

```
Epoch [1/10], Loss: 0.6942412853240967
Epoch [2/10], Loss: 0.6942408084869385
Epoch [3/10], Loss: 0.6942399740219116
Epoch [4/10], Loss: 0.6942387819290161
Epoch [5/10], Loss: 0.6942373514175415
Epoch [6/10], Loss: 0.6942355632781982
Epoch [7/10], Loss: 0.6942337155342102
Epoch [8/10], Loss: 0.6942315697669983
Epoch [9/10], Loss: 0.6942291259765625
Epoch [10/10], Loss: 0.6942267417907715
```

注意：这只是一个简单的例子，在实际应用中的图像分割模型通常更加复杂，并使用更多的层和技术来提高性能和准确度。此外，通常还会使用更大规模的图像数据集进行训练。

6.3.2 DeepLab 语义分割

DeepLab 是一种用于图像语义分割的深度学习模型，它是由 Google 开发的一系列模型，旨在对图像中的每个像素进行语义分类，即将图像分割成不同的语义区域。DeepLab 采用了卷积神经网络和空洞卷积等技术，具有较强的感受野和上下文信息，能够准确地捕捉图像中不同目标的边界和细节。例如，下面是一个使用 DeepLab 实现语义分割的例子。

实例 6-3：使用 DeepLab 实现语义分割

源码路径：daima\6\deep.py

```python
import torch
import torchvision.transforms as transforms
from PIL import Image
import matplotlib.pyplot as plt
from torchvision.models.segmentation import deeplabv3_resnet50

# 加载预训练的 DeepLab 模型
model = deeplabv3_resnet50(pretrained=True)
model.eval()

# 定义图像预处理转换
transform = transforms.Compose([
    transforms.Resize((256, 256)),
    transforms.ToTensor(),
    transforms.Normalize(mean=[0.485, 0.456, 0.406], std=[0.229, 0.224, 0.225])
])
```

```python
# 加载图像
image = Image.open('image.jpg')  # 假设有一张名为 image.jpg 的图像
input_image = transform(image).unsqueeze(0)  # 转换图像并添加批次维度

# 使用模型进行图像语义分割
with torch.no_grad():
    outputs = model(input_image)['out']

# 获取预测结果
predicted_mask = torch.argmax(outputs.squeeze(), dim=0).detach().cpu().numpy()

# 创建预测掩码图像
mask_image = Image.fromarray(predicted_mask.astype('uint8'))

# 创建调色板
palette = [0, 0, 0,      # 背景颜色
           255, 0, 0,    # 人物颜色
           0, 255, 0,    # 车辆颜色
           0, 0, 255]    # 树木颜色
mask_image.putpalette(palette)   # 应用调色板

# 显示原始图像和预测掩码图像
fig, (ax1, ax2) = plt.subplots(1, 2, figsize=(10, 5))
ax1.imshow(image)
ax1.axis('off')
ax1.set_title('Original Image')
ax2.imshow(mask_image)
ax2.axis('off')
ax2.set_title('Predicted Mask')
plt.show()
```

上述代码实现了使用预训练的 DeepLab 模型对图像进行语义分割，并显示原始图像和预测掩码图像。具体实现流程如下：

（1）首先，导入所需的库，包括 torch、torchvision.transforms、PIL 和 matplotlib.pyplot，并导入了 DeepLab 模型 deeplabv3_resnet50。

（2）然后，使用预训练的 DeepLab 模型创建了一个实例，并将其设置为评估模式，即 model.eval()。

（3）定义图像预处理转换 transform，其中包括将图像调整为 256×256 大小、转换为张量和归一化操作。

（4）加载图像，假设图像的文件名为'888.jpg'，并使用定义的转换对图像进行预处理，然后添加一个批次维度，以符合模型的输入要求。

（5）使用 model 对输入图像进行语义分割。在 with torch.no_grad()上下文中，将输入图像传递给模型并获取输出。模型的输出是一个张量，表示每个像素点属于不同类别的概率。

（6）通过在输出张量上使用 torch.argmax 函数，找到每个像素点最可能的类别，并使用.detach().cpu().numpy()将张量转换为 NumPy 数组。

（7）创建预测掩码图像，通过 Image.fromarray 将预测掩码数组转换为 PIL 图像对象。

（8）创建一个调色板 palette，定义每个类别对应的颜色值。然后通过 mask_image.putpalette(palette)方法应用调色板到预测掩码图像。

（9）最后，使用 matplotlib.pyplot 将原始图像和预测掩码图像显示在一个包含两个子图的

图形窗口中，其中左侧子图显示原始图像，右侧子图显示预测掩码图像。

6.4 SSD 目标检测

SSD（single shot multibox detector）是一种常用的目标检测算法，它能够在单个前向传递（single shot）中检测图像中的多个目标（multi-box）。SSD 结合了特征提取网络和多个不同尺度的卷积层，以实现在不同大小的目标上进行检测。

6.4.1 摄像头目标检测

下面实例的功能是使用 SSD（single shot multibox detector）模型实现摄像头图像实时目标检测。

实例 6-4：使用 SSD 模型实现摄像头图像实时目标检测

源码路径： daima\6\fen.py

```python
import cv2
import numpy as np

# 加载预训练的模型和标签
model = cv2.dnn.readNetFromCaffe('deploy.prototxt', 'model.caffemodel')
with open('labels.txt', 'r') as f:
    labels = f.read().splitlines()

# 打开摄像头
cap = cv2.VideoCapture(0)

while True:
    # 读取视频帧
    ret, frame = cap.read()

    # 创建一个blob（二进制大对象）从图像进行前处理
    blob = cv2.dnn.blobFromImage(frame, 0.007843, (300, 300), (127.5, 127.5, 127.5), swapRB=True, crop=False)

    # 将blob输入到模型中进行推理
    model.setInput(blob)
    detections = model.forward()

    # 处理检测结果
    for i in range(detections.shape[2]):
        confidence = detections[0, 0, i, 2]
        if confidence > 0.5:  # 设定置信度阈值为 0.5
            class_id = int(detections[0, 0, i, 1])
            label = labels[class_id]
            x1 = int(detections[0, 0, i, 3] * frame.shape[1])
            y1 = int(detections[0, 0, i, 4] * frame.shape[0])
            x2 = int(detections[0, 0, i, 5] * frame.shape[1])
            y2 = int(detections[0, 0, i, 6] * frame.shape[0])

            # 在帧上绘制检测结果
            cv2.rectangle(frame, (x1, y1), (x2, y2), (0, 255, 0), 2)
```

```
                cv2.putText(frame, label, (x1, y1 - 10), cv2.FONT_HERSHEY_SIMPLEX, 0.9,
(0, 255, 0), 2)

        # 显示帧
        cv2.imshow('SSD Object Detection', frame)

        # 按Q键退出循环
        if cv2.waitKey(1) & 0xFF == ord('q'):
            break

# 释放摄像头和窗口
cap.release()
cv2.destroyAllWindows()
```

对上述代码的具体说明如下:
- 导入所需的库:导入 cv2 用于图像处理和显示,导入 numpy 用于数组操作。
- 加载预训练模型和标签:使用 cv2.dnn.readNetFromCaffe 函数从 deploy.prototxt 和 model.caffemodel 文件中加载预训练的 SSD 模型。同时,从 labels.txt 文件中读取类别标签。
- 打开摄像头:使用 cv2.VideoCapture(0)打开默认的摄像头。
- 进入循环:使用 while 循环来持续读取摄像头的视频帧。
- 创建 blob:使用 cv2.dnn.blobFromImage 函数将当前帧进行预处理,生成一个 blob 对象,作为输入传递给模型。预处理包括尺寸调整、像素值归一化等操作。
- 进行推理:使用模型的 setInput 方法将 blob 输入到模型中进行推理,得到检测结果。通过模型的 forward 方法获取输出。
- 处理检测结果:对每个检测到的目标,获取其置信度、类别 ID、边界框坐标等信息。如果置信度大于阈值(此处设为 0.5),则认为目标检测有效。
- 在帧上绘制检测结果:使用 cv2.rectangle 和 cv2.putText 函数在原始帧上绘制检测到的目标的边界框和类别标签。
- 显示帧:使用 cv2.imshow 函数显示带有检测结果的帧。
- 退出循环:如果按 Q 键,则退出循环。
- 释放资源:释放摄像头并关闭窗口。

6.4.2 基于图像的目标检测

下面的实例的功能是使用 SSD(single shot multibox detector)模型对现有的素材图像实现目标检测。

实例 6-5:使用 SSD 模型对现有的素材图像实现目标检测
源码路径: daima\6\tu.py

```
import cv2

# 加载预训练的模型和标签
model = cv2.dnn.readNetFromCaffe('deploy.prototxt', 'model.caffemodel')
with open('labels.txt', 'r') as f:
    labels = f.read().splitlines()

# 读取图像
```

```python
image = cv2.imread('999.jpg')

# 创建一个blob（二进制大对象）从图像进行前处理
blob = cv2.dnn.blobFromImage(image, 0.007843, (300, 300), (127.5, 127.5, 127.5),
swapRB=True, crop=False)

# 将blob输入到模型中进行推理
model.setInput(blob)
detections = model.forward()

# 处理检测结果
for i in range(detections.shape[2]):
    confidence = detections[0, 0, i, 2]
    if confidence > 0.1:  # 设定置信度阈值为0.5
        class_id = int(detections[0, 0, i, 1])
        label = labels[class_id]
        x1 = int(detections[0, 0, i, 3] * image.shape[1])
        y1 = int(detections[0, 0, i, 4] * image.shape[0])
        x2 = int(detections[0, 0, i, 5] * image.shape[1])
        y2 = int(detections[0, 0, i, 6] * image.shape[0])

        # 在图像上绘制检测结果
        cv2.rectangle(image, (x1, y1), (x2, y2), (0, 255, 0), 2)
        cv2.putText(image, label, (x1, y1 - 10), cv2.FONT_HERSHEY_SIMPLEX, 0.9, (0, 255, 0), 2)

# 显示图像
cv2.imshow('SSD Object Detection', image)
cv2.waitKey(0)
cv2.destroyAllWindows()
```

对上述代码的具体说明如下：

- 使用 cv2.imread()函数读取图像文件，将其存储在 image 变量中。
- 创建一个 blob 对象，对图像进行预处理，包括尺寸调整、像素归一化等。
- 将 blob 对象输入到模型中进行推理，获取目标检测结果。
- 遍历检测结果，提取置信度、类别标签和边界框坐标。
- 根据置信度阈值，筛选出置信度较高的检测结果。
- 在图像上绘制筛选后的检测结果，包括绘制边界框和标签。
- 使用 cv2.imshow()函数显示带有检测结果的图像，并等待按键关闭图像窗口。

执行后将会使用矩形线条标注出图片中的目标区域，如图6-1所示。

图6-1 标注出目标区域

第 7 章 图像分类处理

图像分类是计算机视觉领域中的一个任务，旨在将输入的图像分配到预定义的类别中。它是一种监督学习问题，其中训练数据集包含标注好的图像和相应的类别标签。图像分类在许多实际应用中起着关键作用，如物体识别、图像搜索、人脸识别、医学图像分析等。本章详细讲解使用 Python 语言实现图像分类的知识。

7.1 图像分类介绍

图像分类的关键挑战在于从图像数据中提取有意义的特征，并建立一个能够将这些特征与相应类别关联的模型。随着深度学习方法的兴起，特别是卷积神经网络的应用，图像分类取得了显著的进展。

深度学习模型通过多个卷积层和全连接层来学习图像的特征表示。在卷积层中，模型可以自动学习图像中的边缘、纹理和形状等低级特征。随着网络的深入，高级特征和语义信息也会逐渐被提取出来。通过训练过程中的权重调整，深度学习模型可以自动学习到最能区分不同类别的特征。

图像分类的一般流程如下：

（1）数据收集和准备：收集适当的图像数据集，并将其划分为训练集、验证集和测试集。图像数据集应包含各个类别的典型样本，并确保标注准确和完整。

（2）特征提取和预处理：将原始图像转换为可用于分类的特征向量。传统方法中常用的特征提取方法包括边缘检测、颜色直方图、纹理特征等。在深度学习中，使用卷积神经网络可以自动学习特征表示，而无须手动设计特征。

（3）模型选择和训练：选择适合任务的图像分类模型，如 LeNet、AlexNet、VGGNet、ResNet、Inception 等。通过使用训练集对模型进行训练，通过梯度下降等优化算法来调整模型的权重和参数，使其能够准确地分类图像。

（4）模型评估和调优：使用验证集对训练得到的模型进行评估，计算分类准确率、精确率、召回率等指标。根据评估结果，可以调整模型的超参数、模型结构或采用正则化等技术来提高模型性能。

（5）模型测试和应用：使用测试集对经过调优的模型进行最终评估，评估模型的泛化能力和性能。在实际应用中，将训练好的模型应用于新的图像数据，进行分类预测。

图像分类是计算机视觉领域的一个广泛研究领域，不断涌现出新的方法和技术，如迁移学习、多尺度处理、注意力机制等，以提高图像分类的准确性和鲁棒性。

7.2 基于特征提取和机器学习的图像分类

传统的图像分类方法通常涉及手工设计的特征提取步骤，例如边缘检测、纹理特征提取、颜色直方图等。这些提取的特征被输入到机器学习算法中,如支持向量机(SVM)、随机森林(random forest)和 K-近邻(K-nearest neighbors)等进行分类。

7.2.1 基本流程

使用特征提取和机器学习方法实现图像分类的基本流程如下:

(1)数据准备:首先,需要收集并准备好带有标注的图像数据集。确保数据集中包含各个类别的典型样本,并确保标注准确和完整。

(2)特征提取:特征提取是将图像转换为机器学习算法能够理解和处理的特征向量的过程。常见的特征提取方法包括:

- 边缘检测:使用边缘检测算法(如 sobel、canny 等)提取图像的边缘信息。
- 颜色直方图:将图像中各个颜色通道的像素值统计为直方图,用于表示颜色分布。
- 纹理特征:提取图像中的纹理信息,如灰度共生矩阵(GLCM)和局部二值模式(LBP)等。
- 尺度不变特征变换(SIFT):提取具有尺度和旋转不变性的局部特征描述子。
- 主成分分析(PCA):将高维的图像特征降维到低维空间中,以减少特征的维度和冗余信息。

(3)特征表示:将提取得到的特征转换为机器学习算法所需的向量形式。可以将特征向量简单地按照一定的规则进行拼接或者使用降维方法(如 PCA)将其转换为较低维度的特征表示。

(4)数据划分:将准备好的特征向量及对应的标签划分为训练集和测试集,通常采用交叉验证等方法进行划分。

(5)机器学习模型训练和分类:选择适当的机器学习算法,如支持向量机、随机森林或 K-近邻等。使用训练集的特征向量和标签进行模型训练,通过优化算法调整模型参数。训练完成后,使用测试集的特征向量进行分类预测,并评估分类准确率、精确率、召回率等指标。

(6)模型评估和调优:根据评估结果,可以调整机器学习模型的超参数、特征选择、模型选择等来提高分类性能。常用的方法包括网格搜索、交叉验证等。

需要注意的是,特征提取和机器学习方法在处理大规模图像数据集时可能存在一些限制,例如手工设计特征可能无法捕捉到复杂的语义信息。然而,在小规模数据集或资源受限的环境中,特征提取和机器学习方法仍然是一个有效的选择。随着深度学习技术的发展,深度学习方法在图像分类任务上取得了更好的性能,但传统方法仍然具有一定的实用性和应用场景。

7.2.2 基于 scikit-learn 机器学习的图像分类

机器学习从开始到建模的基本流程是:获取数据、数据预处理、训练模型、模型评估、预测、分类。本节我们将根据传统机器学习的流程,来介绍在每一步流程中都有哪些常用的函数以及它们的用法是怎么样的。例如,下面是一个使用库 scikit-learn 实现图像分类的例子,其中采用了特征提取和机器学习方法。

实例 7-1:使用库 scikit-learn 实现图像分类

源码路径: daima\7\catdog.py

```
import os
import cv2
```

```python
import numpy as np
from sklearn.model_selection import train_test_split
from sklearn.svm import SVC
from sklearn.metrics import accuracy_score

# 数据集路径和类别标签
dataset_path = "path_to_dataset"
categories = ["cat", "dog"]

# 提取特征的函数(示例中使用颜色直方图作为特征)
def extract_features(image_path):
    image = cv2.imread(image_path)
    image = cv2.resize(image, (100, 100))  # 调整图像大小
    hist = cv2.calcHist([image], [0, 1, 2], None, [8, 8, 8], [0, 256, 0, 256, 0, 256])
# 计算颜色直方图
    hist = cv2.normalize(hist, hist).flatten()  # 归一化并展平
    return hist

# 加载图像数据和标签
data = []
labels = []
for category in categories:
    category_path = os.path.join(dataset_path, category)
    for image_name in os.listdir(category_path):
        image_path = os.path.join(category_path, image_name)
        features = extract_features(image_path)
        data.append(features)
        labels.append(category)

# 将数据集划分为训练集和测试集
X_train, X_test, y_train, y_test = train_test_split(data, labels, test_size=0.2, random_state=42)

# 使用支持向量机(SVM)作为分类器
classifier = SVC()
classifier.fit(X_train, y_train)

# 在测试集上进行预测
y_pred = classifier.predict(X_test)

# 计算分类准确率
accuracy = accuracy_score(y_test, y_pred)
print("Accuracy:", accuracy)
```

在上述代码中,首先定义了数据集路径和类别标签。然后,通过函数 extract_features()提取图像的特征,这里使用的是颜色直方图。接下来,遍历数据集中的图像,提取特征并将其添加到数据列表中,同时记录对应的类别标签。使用函数 train_test_split()将数据集划分为训练集和测试集。在训练阶段,使用 SVC 类作为分类器,通过调用 fit 方法对训练数据进行训练。最后,使用训练好的分类器在测试集上进行预测,并计算分类准确率。执行后会输出:

Accuracy: 0.7857142857142857

注意:上面输出的准确率只有 0.78 的原因是作者在运行时使用的数据集过小导致的,机器学习算法通常需要足够的数据来学习和泛化。在源码中提供了 kaggle dog VS.cat 数据集(下载地址:https://aistudio.baidu.com/aistudio/datasetdetail/11544),在这个训练集中共有 25 000

张图片，猫狗各一半。格式为 dog.xxx.jpg/cat.xxx.jpg（xxx 为编号）测试集 12 500 张，没标定是猫还是狗，格式为 xxx.jpg。大家可以使用这个数据集进行训练识别，准确率会大大提高。这只是一个简单的示例，实际应用中可能需要根据具体情况进行更多的数据预处理、特征选择、模型调优等步骤。另外，可以尝试使用其他特征提取方法和机器学习算法，以及结合交叉验证等技术来提高分类性能。

再看下面的实例文件 hua.py，实现了鸢尾花识别的功能。这是一个经典的机器学习分类问题，它的数据样本中包括了 4 个特征变量，1 个类别变量，样本总数为 150。本实例的目标是根据花萼长度（sepal length）、花萼宽度（sepal width）、花瓣长度（petal length）、花瓣宽度（petal width）这四个特征来识别出鸢尾花属于山鸢尾（iris-setosa）、变色鸢尾（iris-versicolor）和维吉尼亚鸢尾（iris-virginica）中的哪一种。

实例 7-2：实现鸢尾花的识别功能

源码路径：daima\7\hua.py

```python
# 引入数据集，sklearn 包含众多数据集
from sklearn import datasets
# 将数据分为测试集和训练集
from sklearn.model_selection import train_test_split
# 利用邻近点方式训练数据
from sklearn.neighbors import KNeighborsClassifier

# 引入数据，本次导入鸢尾花数据，iris 数据包含 4 个特征变量
iris = datasets.load_iris()
# 特征变量
iris_X = iris.data
# print(iris_X)
print('特征变量的长度', len(iris_X))
# 目标值
iris_y = iris.target
print('鸢尾花的目标值', iris_y)
# 利用 train_test_split 进行训练集和测试机进行分开，test_size 占 30%
X_train, X_test, y_train, y_test = train_test_split(iris_X, iris_y, test_size=0.3)
# 我们看到训练数据的特征值分为 3 类
print(y_train)

# 训练数据
# 引入训练方法
knn = KNeighborsClassifier()
# 进行填充测试数据进行训练
knn.fit(X_train, y_train)

params = knn.get_params()
print(params)
score = knn.score(X_test, y_test)
print("预测得分为: %s" % score)

# 预测数据，预测特征值
print(knn.predict(X_test))

# 打印真实特征值
print(y_test)
```

执行后会输出训练和预测结果：

特征变量的长度 150

```
鸢尾花的目标值 [0 0 0 0 0 0 0 0 0 0 0 0 0 0 0 0 0 0 0 0 0 0 0 0 0 0 0 0 0 0
 0 0
 0 0 0 0 0 0 0 0 0 0 0 0 0 0 0 1 1 1 1 1 1 1 1 1 1 1 1 1 1 1 1 1 1 1 1
 1 1 1 1 1 1 1 1 1 1 1 1 1 1 1 1 1 1 1 1 1 1 1 1 1 2 2 2 2 2 2 2 2 2 2
 2 2 2 2 2 2 2 2 2 2 2 2 2 2 2 2 2 2 2 2 2 2 2 2 2 2 2 2 2 2 2 2 2 2 2
 2 2]
[2 1 2 1 0 2 0 1 0 1 0 1 1 2 1 0 0 0 1 2 2 2 1 1 2 1 0 2 0 0 2 2 2 0
 1 1 2 0 2 1 1 1 2 0 0 1 1 1 1 0 1 0 2 2 2 1 1 0 0 2 0 2 1 0 2 1 1 0 2 2
 2 0 1 1 0 2 0 1 2 2 1 1 1 0 1 1 2 0 0 2 0 0 1 2 0 0 0 1 2 2]
{'algorithm': 'auto', 'leaf_size': 30, 'metric': 'minkowski', 'metric_params': None,
'n_jobs': None, 'n_neighbors': 5, 'p': 2, 'weights': 'uniform'}
预测得分为: 1.0
[0 2 0 0 1 0 1 1 0 0 2 2 1 0 2 2 1 0 0 2 0 2 1 0 2 1 2 2 2 2 0 2 0 0 1 2 2
 0 1 2 1 1 1 0 1]
[0 2 0 0 1 0 1 1 0 0 2 2 1 0 2 2 1 0 0 2 0 2 1 0 2 1 2 2 2 2 0 2 0 0 1 2 2
 0 1 2 1 1 1 0 1]
```

7.2.3 分类算法

下面的实例文件 fen.py，功能是绘制不同分类器的分类概率。我们使用一个 3 类的数据集，并使用支持向量分类器、带 L1 和 L2 惩罚项的 Logistic 回归，使用 One-Vs-Rest 或多项设置以及高斯过程分类对其进行分类。在默认情况下，线性 SVC 不是概率分类器，但在本例中它有一个内建校准选项（probability=True）。箱外的 One-Vs-Rest 的逻辑回归不是一个多分类的分类器，因此，与其他估计器相比，它在分离第 2 类和第 3 类时有更大的困难。

实例 7-3：绘制不同分类器的分类概率

源码路径： daima\7\fen.py

```python
import matplotlib.pyplot as plt
import numpy as np

from sklearn.metrics import accuracy_score
from sklearn.linear_model import LogisticRegression
from sklearn.svm import SVC
from sklearn.gaussian_process import GaussianProcessClassifier
from sklearn.gaussian_process.kernels import RBF
from sklearn import datasets

iris = datasets.load_iris()
X = iris.data[:, 0:2]  # we only take the first two features for visualization
y = iris.target

n_features = X.shape[1]

C = 10
kernel = 1.0 * RBF([1.0, 1.0])  # for GPC

# Create different classifiers.
classifiers = {
    'L1 logistic': LogisticRegression(C=C, penalty='l1',
                                      solver='saga',
                                      multi_class='multinomial',
                                      max_iter=10000),
    'L2 logistic (Multinomial)': LogisticRegression(C=C, penalty='l2',
                                                    solver='saga',
                                                    multi_class='multinomial',
```

```python
                                    max_iter=10000),
    'L2 logistic (OvR)': LogisticRegression(C=C, penalty='l2',
                                    solver='saga',
                                    multi_class='ovr',
                                    max_iter=10000),
    'Linear SVC': SVC(kernel='linear', C=C, probability=True,
                      random_state=0),
    'GPC': GaussianProcessClassifier(kernel)
}

n_classifiers = len(classifiers)

plt.figure(figsize=(3 * 2, n_classifiers * 2))
plt.subplots_adjust(bottom=.2, top=.95)

xx = np.linspace(3, 9, 100)
yy = np.linspace(1, 5, 100).T
xx, yy = np.meshgrid(xx, yy)
Xfull = np.c_[xx.ravel(), yy.ravel()]

for index, (name, classifier) in enumerate(classifiers.items()):
    classifier.fit(X, y)

    y_pred = classifier.predict(X)
    accuracy = accuracy_score(y, y_pred)
    print("Accuracy (train) for %s: %0.1f%% " % (name, accuracy * 100))

    # View probabilities:
    probas = classifier.predict_proba(Xfull)
    n_classes = np.unique(y_pred).size
    for k in range(n_classes):
        plt.subplot(n_classifiers, n_classes, index * n_classes + k + 1)
        plt.title("Class %d" % k)
        if k == 0:
            plt.ylabel(name)
        imshow_handle = plt.imshow(probas[:, k].reshape((100, 100)),
                                   extent=(3, 9, 1, 5), origin='lower')
        plt.xticks(())
        plt.yticks(())
        idx = (y_pred == k)
        if idx.any():
            plt.scatter(X[idx, 0], X[idx, 1], marker='o', c='w', edgecolor='k')

ax = plt.axes([0.15, 0.04, 0.7, 0.05])
plt.title("Probability")
plt.colorbar(imshow_handle, cax=ax, orientation='horizontal')

plt.show()
```

上述代码实现了在一个二维特征空间中可视化多个分类器的决策边界和分类概率。这段代码主要用于演示不同分类器在二维特征空间中的分类效果和概率分布情况。通过观察决策边界和概率图像，可以了解不同分类器在鸢尾花数据集上的性能和特点。具体实现流程如下：

（1）首先，导入所需的库和模块。

（2）然后，加载鸢尾花（Iris）数据集，其中 X 是特征矩阵，包含了鸢尾花的萼片长度和宽度两个特征；y 是目标变量，包含了鸢尾花的类别标签。

（3）接下来，定义了一个正态核函数（RBF）作为高斯过程分类器（GPC）的核函数，并设置了惩罚参数 C。

（4）创建不同的分类器，包括 L1 正则化的逻辑回归（L1 logistic）、L2 正则化的逻辑回归（L2 logistic）、线性支持向量机（Linear SVC）和高斯过程分类器（GPC）。这些分类器使用不同的参数和算法来进行训练和分类。

（5）使用 plt.subplots_adjust 调整子图的位置和间距，创建一个绘图窗口，并设置图像的大小。

（6）生成了一个网格点坐标矩阵（Xfull），用于在整个特征空间中生成预测结果。

（7）对于每个分类器，依次进行训练和预测，计算训练集上的准确率，并打印出来。

（8）对于每个分类器，绘制类别的预测概率图像，显示每个类别的概率分布情况。

（9）通过调整颜色条的位置和大小，将颜色条添加到图像中。

（10）最后使用 plt.show()显示绘制的图像。

执行后会输出下面的结果，并在 Matplotlib 中绘制三种分类的概率，如图 7-1 所示。

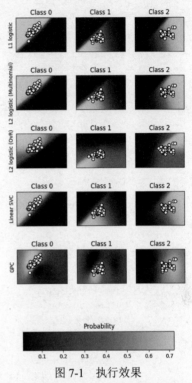

图 7-1 执行效果

```
Accuracy (train) for L1 logistic: 83.3%
Accuracy (train) for L2 logistic (Multinomial): 82.7%
Accuracy (train) for L2 logistic (OvR): 79.3%
Accuracy (train) for Linear SVC: 82.0%
Accuracy (train) for GPC: 82.7%
```

7.2.4 聚类算法

下面的实例文件 face.py 的功能是使用一个大型的 Faces 数据集学习一组组成面部的 20×20 的图像修补程序。本实例展示了使用 scikit-learn 在线 API 学习按块处理一个大型数据集的方法。本实例处理的方法是一次加载一个图像，并从这个图像中随机提取 50 个补丁。一旦积累了 500 个补丁（使用 10 个图像），则运行在线 KMeans 对象 MiniBatchKMeans 的 partial_fit 方法。在连续调用 partial-fit 期间，某些聚类会被重新分配。这是因为它们所代表的补丁数量太少了，所以最好选择一个随机的新聚类。

实例 7-4：使用一个 Faces 数据集学习一组组成面部的 20×20 的图像修补程序

源码路径：daima\7\face.py

```
import time

import matplotlib.pyplot as plt
import numpy as np
```

```python
from sklearn import datasets
from sklearn.cluster import MiniBatchKMeans
from sklearn.feature_extraction.image import extract_patches_2d

faces = datasets.fetch_olivetti_faces()

# #############################################################################
# Learn the dictionary of images

print('Learning the dictionary... ')
rng = np.random.RandomState(0)
kmeans = MiniBatchKMeans(n_clusters=81, random_state=rng, verbose=True)
patch_size = (20, 20)

buffer = []
t0 = time.time()

# 在整个数据集上循环6次
index = 0
for _ in range(6):
    for img in faces.images:
        data = extract_patches_2d(img, patch_size, max_patches=50,
                                  random_state=rng)
        data = np.reshape(data, (len(data), -1))
        buffer.append(data)
        index += 1
        if index % 10 == 0:
            data = np.concatenate(buffer, axis=0)
            data -= np.mean(data, axis=0)
            data /= np.std(data, axis=0)
            kmeans.partial_fit(data)
            buffer = []
        if index % 100 == 0:
            print('Partial fit of %4i out of %i'
                  % (index, 6 * len(faces.images)))

dt = time.time() - t0
print('done in %.2fs.' % dt)

# #############################################################################
# Plot the results
plt.figure(figsize=(4.2, 4))
for i, patch in enumerate(kmeans.cluster_centers_):
    plt.subplot(9, 9, i + 1)
    plt.imshow(patch.reshape(patch_size), cmap=plt.cm.gray,
               interpolation='nearest')
    plt.xticks(())
    plt.yticks(())


plt.suptitle('Patches of faces\nTrain time %.1fs on %d patches' %
             (dt, 8 * len(faces.images)), fontsize=16)
plt.subplots_adjust(0.08, 0.02, 0.92, 0.85, 0.08, 0.23)

plt.show()
```

上述代码主要用于学习和展示图像字典中的图像块。通过将图像块提取为数据，然后使用MiniBatchKMeans算法对数据进行聚类，可以学习到一组具有代表性的图像块，用于后续的图像处理和特征表示。本实例实现了使用 MiniBatchKMeans 算法学习图像字典，并展示学习到的图像字典中的图像块。具体实现流程如下：

（1）首先，导入需要的库和模块，包括 time、matplotlib.pyplot 和 numpy。

（2）通过 datasets.fetch_olivetti_faces()加载了奥利维蒂人脸数据集（olivetti faces），该数据集包含了一组人脸图像。

（3）接下来，定义一个 MiniBatchKMeans 聚类器，并设置了聚类的数量 n_clusters 为 81，以及随机数生成器的种子 random_state。

（4）定义一个图像块的大小 patch_size，它在本例中是一个 20×20 的矩形。

（5）创建一个空列表 buffer，用于存储图像块的数据。

（6）通过循环遍历整个数据集 6 次，对每张人脸图像进行处理。

（7）在每次循环中，使用 extract_patches_2d 函数从图像中提取图像块，设置最大提取数量为 50，然后将图像块的数据进行重塑和规范化处理，将其添加到 buffer 列表中。

（8）每当 buffer 列表中的图像块数量达到 10 个时，将它们连接成一个数据矩阵，并进行均值归一化处理。

（9）使用 kmeans.partial_fit 对数据进行部分拟合（partial fit）来更新聚类器的参数。

（10）每当处理了 100 个图像块时，打印出部分拟合的进度。

（11）通过计算总共花费的时间来评估学习过程的耗时。

（12）使用 matplotlib.pyplot 绘制了学习到的图像字典中的图像块。循环遍历聚类器的聚类中心，并使用 plt.subplot 在子图中显示图像块。设置合适的标题和调整子图的布局。

（13）通过 plt.show()显示绘制的图像。

代码执行后的效果如图 7-2 所示。

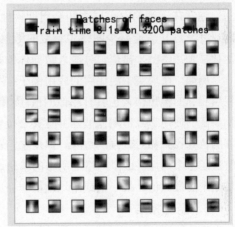

图 7-2　组成面部的图像修补效果

7.3　基于卷积神经网络的图像分类

前面介绍的是基于机器学习的图像分类，接下来介绍基于深度学习的图像分类，基于深度学习的图像分类技术主要有卷积神经网络、迁移学习、循环神经网络、卷积循环神经网络等。本节讲解基于卷积神经网络的图像分类知识。

7.3.1　卷积神经网络基本结构

基于卷积神经网络（CNN）是当前最主流和成功的图像分类方法之一。它通过多个卷积层和池化层来学习图像的特征表示，随后通过全连接层进行分类。著名的 CNN 模型包括 LeNet、AlexNet、VGGNet、ResNet、Inception 等。基础的 CNN 由卷积（convolution）、激活（activation）和池化（pooling）三种结构组成。CNN 输出的结果是每幅图像的特定特征空间。当处理图像分类任务时，会把CNN输出的特征空间作为全连接层或全连接神经网络（fully connected neural network，FCN）的输入，用全连接层来完成从输入图像到标签集的映射，即分类。当然，整个过程最重要的工作就是如何通过训练数据迭代调整网络权重，也就是后向传播算法。

下面，详细讲解卷积神经网络的基本结构。

1. 卷积层

卷积层是卷积网络的核心，大多数计算都是在卷积层中进行的。卷积层的功能是实现特征提取，卷积网络的参数是由一系列可以学习的滤波器集合构成的，每个滤波器在宽度和高度上都比较小，但是深度输入和数据保持一致。当滤波器沿着图像的宽和高滑动时，会生成一个二维的激活图。

卷积层的参数是由一些可学习的滤波器集合构成的。每个滤波器在空间上（宽度和高度）都比较小，但是深度和输入数据一致（这一点很重要，后面会具体介绍）。直观地来说，网络会让滤波器学习到当它看到某些类型的视觉特征时就激活，具体的视觉特征可能是某些方位上的边界，或者在第一层上某些颜色的斑点，甚至可以是网络更高层上的蜂巢状或者车轮状图案。

2. 池化层

通常在连续的卷积层之间会周期性地插入一个池化层，它的作用是逐渐降低数据体的空间尺寸，这样就能减少网络中参数的数量，使得计算资源耗费变少，也能有效控制过拟合。池化层使用MAX操作，对输入数据体的每一个深度切片独立进行操作，改变它的空间尺寸。

例如，在现实中池化层的应用例子：图像中的相邻像素倾向于具有相似的值，因此通常卷积层相邻的输出像素也具有相似的值。这意味着，卷积层输出中包含的大部分信息都是冗余的。如果我们使用边缘检测滤波器并在某个位置找到强边缘，那么也可能会在距离这个像素1个偏移的位置找到相对较强的边缘。但是它们都一样是边缘，我们并没有找到任何新东西。池化层解决了这个问题。这个网络层所做的就是通过减小输入的大小降低输出值的数量。池化一般通过简单的最大值、最小值或平均值操作完成。

3. 全连接层

全连接层的输入层是前面的特征图，会将特征图中所有的神经元变成全连接的样子。这个过程为了防止过拟合会引入Dropout。在进入全连接层之前，使用全局平均池化能够有效地降低过拟合。

对于每个卷积层来说，都存在一个可以实现相同前向传播函数的全连接层。这个全连接层的权重矩阵是巨大的，除了某些特定的块（感受野）之外，其余部分都是零。在非零部分中，大部分元素都是相等的（权值共享）。这意味着全连接层的权重矩阵中，除了与感受野对应的位置有非零值外，其他位置都是零。

需要注意的是，要区分"将全连接层转换为卷积层"和"使用矩阵乘法实现卷积"两者的不同。后者实际上仍然在计算卷积，只是将其展开为矩阵相乘的形式，而不是"将全连接层转换为卷积层"。因此，除非权重本身就是零，否则在使用矩阵乘法实现卷积的过程中不会出现零值权重。

4. 激活层

激活层也被称为激活函数（activation function），是在人工神经网络的神经元上运行的函数，负责将神经元的输入映射到输出端。激活层对于人工神经网络模型去学习、理解非常复杂和非线性的函数来说具有十分重要的作用。它们将非线性特性引入到我们的网络中。例如，在矩阵运算应用中，在神经元中输入的 inputs 通过加权求和后，还被作用于一个函数，这个函数就是激活函数。引入激活函数是为了增加神经网络模型的非线性特征。在神经网络中，每一层都包含线性变换（例如矩阵相乘），如果没有激活函数，多个线性层的叠加仍然只能表示线性关系，无法捕捉复杂的非线性关系。

5. Dropout 层

Dropout 是指深度学习训练过程中，对神经网络训练单元按照一定的概率将其从网络中移除，注意是暂时，对于随机梯度下降来说，由于是随机丢弃，故而每一个 mini-batch 都在训练不同的网络。

Dropout 的作用是在训练神经网络模型时由于样本数据过少，为防止过拟合而采用的技术。首先，想象我们现在只训练一个特定的网络，当迭代次数增多的时候，可能出现网络对训练集拟合得很好（在训练集上 loss 很小），但是对验证集的拟合程度很差的情况。所以有了这样的想法：可不可以让每次跌代随机更新网络参数（weights），引入这样的随机性就可以增加网络的概括能力，所以就有了 Dropout。

在训练的时候，我们只需要按一定的概率（retaining probability）p 来对 Weight 层的参数进行随机采样，将这个子网络作为此次更新的目标网络。可以想象，如果整个网络有 n 个参数，那么我们可用的子网络个数为 2^n。并且当 n 很大时，每次迭代更新使用的子网络基本上不会重复，从而避免了某一个网络被过分拟合到训练集上。

那么在测试的时候怎么办呢？一种基础的方法是把 2^n 个子网络都用来做测试，然后以某种投票机制将所有结果结合一下（比如说平均一下），然后得到最终的结果。但是，由于 n 实在是太大了，这种方法实际中完全不可行。所以有人提出做一个大致的估计即可，从 2^n 个网络中随机选取 m 个网络做测试，最后再用某种投票机制得到最终的预测结果。这种想法当然可行，当 m 很大但又远小于 2^n 时，能够很好地逼近原 2^n 个网络结合起来的预测结果。但是还有更好的办法：那就是 dropout 自带的功能，能够通过一次测试得到逼近于原 2^n 个网络组合起来的预测能力。

6. BN 层

BN 的全称为 Batch Normalization，在进行深度网络训练时，大都会采取这种算法。尽管梯度下降法训练神经网络很简单高效，但是需要人为地去选择参数，比如学习率、参数初始化、权重衰减系数、Dropout 比例等，而且这些参数的选择对于训练结果至关重要，以至于我们很多时间都浪费到这些参数的调整上。BN 算法的强大之处在下面几个方面：

- 可以选择较大的学习率，使得训练速度增长很快，具有快速收敛性。
- 可以不去理会 Dropout，L2 正则项参数的选择，如果选择使用 BN，甚至可以去掉这两项。
- 去掉局部响应归一化层。（AlexNet 中使用的方法，BN 层出来之后这个就不再用了）。
- 可以把训练数据打乱，防止每批训练的时候，某一个样本被经常挑选到。

首先，来说归一化的问题，神经网络训练开始前，都要对数据做一个归一化处理，归一化有很多好处，原因是网络学习的过程的本质就是学习数据分布，一旦训练数据和测试数据的分布不同，那么网络的泛化能力就会大大降低；另一方面，每一批次的数据分布如果不相同的话，那么网络就要在每次迭代的时候都去适应不同的分布，这样会大大降低网络的训练速度，这也就是为什么要对数据做归一化预处理的原因。另外，对图片进行归一化处理还可以处理光照、对比度等影响。例如，网络一旦训练起来，参数就要发生更新，除了输入层的数据外，其他层的数据分布是一直发生变化的，因为在训练的时候，网络参数的变化就会导致后面输入数据的分布变化。比如第二层输入，是由输入数据和第一层参数得到的，而第一层的参数随着训练一直变化，势必会引起第二层输入分布的改变，把这种改变称为 internal covariate shift，BN 就是为了解决这个问题而诞生的。

经过上面的描述可以得出一个结论：卷积神经网络主要由这几类层构成：输入层、卷积层、激活层层、池化（pooling）层和全连接层（全连接层和常规神经网络中的一样）。通过将这些层叠加起来，就可以构建一个完整的卷积神经网络。在实际应用中往往将卷积层与激活层共同称为卷积层，所以卷积层经过卷积操作也是要经过激活函数的。具体说来，卷积层和全连接层（CONV/FC）对输入执行变换操作的时候，不仅会用到激活函数，还会用到很多参数，即神经元的权值 w 和偏差 b；而激活层和池化层则是进行一个固定不变的函数操作。卷积层和全连接层中的参数会随着梯度下降被训练，这样卷积神经网络计算出的分类评分就能和训练集中的每个图像的标签吻合了。

7.3.2　第一个卷积神经网络程序

在下面的实例文件 cnn01.py 中，将使用 TensorFlow 创建一个卷积神经网络模型，并可视化评估这个模型。

实例 7-5：创建一个卷积神经网络模型并可视化评估

文件 cnn01.py 的具体实现流程如下：

（1）导入 TensorFlow 模块，代码如下：

```
import tensorflow as tf
from tensorflow.keras import datasets, layers, models
import matplotlib.pyplot as plt
```

（2）下载并准备 CIFAR10 数据集。

CIFAR10 数据集包含 10 类，共 60 000 张彩色图片，每类图片有 6 000 张。此数据集中 50 000 个样例被作为训练集，剩余 10 000 个样例作为测试集。类之间相互独立，不存在重叠的部分。代码如下：

```
(train_images, train_labels), (test_images, test_labels) = datasets.cifar10.load_data()

# 将像素的值标准化至0~1的区间内
train_images, test_images = train_images / 255.0, test_images / 255.0
```

（3）验证数据。

将数据集中的前 25 张图片和类名打印出来，确保数据集被正确加载。代码如下：

```
class_names = ['airplane', 'automobile', 'bird', 'cat', 'deer',
               'dog', 'frog', 'horse', 'ship', 'truck']
plt.figure(figsize=(10,10))
for i in range(25):
    plt.subplot(5,5,i+1)
    plt.xticks([])
    plt.yticks([])
```

```
    plt.grid(False)
    plt.imshow(train_images[i], cmap=plt.cm.binary)
    # 由于 CIFAR 的标签是 array，
    # 因此你需要额外的索引（index）
    plt.xlabel(class_names[train_labels[i][0]])
plt.show()
```

执行后将可视化显示数据集中的前 25 张图片和类名，如图 7-3 所示。

图 7-3 可视化显示数据集中的前 25 张图片和类名

（4）构造卷积神经网络模型。

通过如下代码声明了一个常见卷积神经网络，由几个 Conv2D 和 MaxPooling2D 层组成。

```
model = models.Sequential()
model.add(layers.Conv2D(32, (3, 3), activation='relu', input_shape=(32, 32, 3)))
model.add(layers.MaxPooling2D((2, 2)))
model.add(layers.Conv2D(64, (3, 3), activation='relu'))
model.add(layers.MaxPooling2D((2, 2)))
model.add(layers.Conv2D(64, (3, 3), activation='relu'))
```

CNN 的输入是张量（Tensor）形式的（image_height, image_width, color_channels），包含了图像高度、宽度及颜色信息。不需要输入 batch size。如果你不熟悉图像处理，颜色信息建议使用 RGB 色彩模式，此模式下，color_channels 为(R,G,B)分别对应 RGB 的三个颜色通道（color channel）。在此示例中，我们的 CNN 输入，CIFAR 数据集中的图片，形状是(32, 32, 3)。可以在声明第一层时将形状赋值给参数 input_shape。声明 CNN 结构的代码是：
```
model.summary()
```
执行后会输出显示模型的基本信息：
```
Model: "sequential"
_____
Layer (type)                 Output Shape              Param #
=================================================================
conv2d (Conv2D)              (None, 30, 30, 32)        896
_____
max_pooling2d (MaxPooling2D) (None, 15, 15, 32)        0
_____
conv2d_1 (Conv2D)            (None, 13, 13, 64)        18496
_____
max_pooling2d_1 (MaxPooling2 (None, 6, 6, 64)          0
_____
conv2d_2 (Conv2D)            (None, 4, 4, 64)          36928
=================================================================
Total params: 56,320
Trainable params: 56,320
Non-trainable params: 0
```

在执行后输出显示的结构中可以看到，每个 Conv2D 和 MaxPooling2D 层的输出都是一个三维的张量（tensor），其形状描述了（height, width, channels）。越深的层中，宽度和高度都会收缩。每个 Conv2D 层输出的通道数量（channels）取决于声明层时的第一个参数（如：上面代码中的 32 或 64）。这样，由于宽度和高度的收缩，您便可以（从运算的角度）增加每个 Conv2D 层输出的通道数量（channels）。

（5）增加 Dense 层。

Dense 层等同于全连接层，在模型的最后，将把卷积后的输出张量[本例中形状为(4, 4, 64)]传给一个或多个 Dense 层来完成分类。Dense 层的输入为向量（一维），但前面层的输出是三维的张量。因此需要将三维张量展开（flatten）到一维，之后再传入一个或多个 Dense 层。CIFAR 数据集有 10 个类，因此最终的 Dense 层需要 10 个输出及一个 softmax 激活函数。代码如下：
```
model.add(layers.Flatten())
model.add(layers.Dense(64, activation='relu'))
model.add(layers.Dense(10))
```
此时通过如下代码查看完整 CNN 的结构：
```
model.summary()
```
执行后会输出显示：
```
Model: "sequential"
_____
Layer (type)                 Output Shape              Param #
=================================================================
conv2d (Conv2D)              (None, 30, 30, 32)        896
_____
max_pooling2d (MaxPooling2D) (None, 15, 15, 32)        0
_____
conv2d_1 (Conv2D)            (None, 13, 13, 64)        18496
_____
max_pooling2d_1 (MaxPooling2 (None, 6, 6, 64)          0
_____
conv2d_2 (Conv2D)            (None, 4, 4, 64)          36928
```

```
flatten (Flatten)              (None, 1024)              0
_____
dense (Dense)                  (None, 64)                65600
_____
dense_1 (Dense)                (None, 10)                650
=================================================================
```

由此可以看出,在被传入两个Dense层之前,形状为(4, 4, 64)的输出被展平成了形状为(1024)的向量。

(6) 编译并训练模型,代码如下:

```
model.compile(optimizer='adam',
              loss=tf.keras.losses.SparseCategoricalCrossentropy(from_logits=True),
              metrics=['accuracy'])

history = model.fit(train_images, train_labels, epochs=10,
                    validation_data=(test_images, test_labels))
```

执行后会输出显示训练过程:

```
Epoch 1/10
1563/1563 [==============================] - 7s 3ms/step - loss: 1.5216 - accuracy: 0.4446 - val_loss: 1.2293 - val_accuracy: 0.5562
Epoch 2/10
1563/1563 [==============================] - 5s 3ms/step - loss: 1.1654 - accuracy: 0.5857 - val_loss: 1.0774 - val_accuracy: 0.6143
Epoch 3/10
1563/1563 [==============================] - 5s 3ms/step - loss: 1.0172 - accuracy: 0.6460 - val_loss: 1.0041 - val_accuracy: 0.6399
Epoch 4/10
1563/1563 [==============================] - 5s 3ms/step - loss: 0.9198 - accuracy: 0.6795 - val_loss: 0.9946 - val_accuracy: 0.6540
Epoch 5/10
1563/1563 [==============================] - 5s 3ms/step - loss: 0.8449 - accuracy: 0.7060 - val_loss: 0.9169 - val_accuracy: 0.6792
Epoch 6/10
1563/1563 [==============================] - 5s 3ms/step - loss: 0.7826 - accuracy: 0.7264 - val_loss: 0.8903 - val_accuracy: 0.6922
Epoch 7/10
1563/1563 [==============================] - 5s 3ms/step - loss: 0.7338 - accuracy: 0.7441 - val_loss: 0.9217 - val_accuracy: 0.6879
Epoch 8/10
1563/1563 [==============================] - 5s 3ms/step - loss: 0.6917 - accuracy: 0.7566 - val_loss: 0.8799 - val_accuracy: 0.6990
Epoch 9/10
1563/1563 [==============================] - 5s 3ms/step - loss: 0.6431 - accuracy: 0.7740 - val_loss: 0.9013 - val_accuracy: 0.6982
Epoch 10/10
1563/1563 [==============================] - 5s 3ms/step - loss: 0.6074 - accuracy: 0.7882 - val_loss: 0.8949 - val_accuracy: 0.7075
```

(7) 评估我们在上面实现的卷积神经网络模型,首先可视化展示评估过程,代码如下:

```
plt.plot(history.history['accuracy'], label='accuracy')
plt.plot(history.history['val_accuracy'], label = 'val_accuracy')
plt.xlabel('Epoch')
plt.ylabel('Accuracy')
plt.ylim([0.5, 1])
plt.legend(loc='lower right')
plt.show()
```

```
test_loss, test_acc = model.evaluate(test_images,  test_labels, verbose=2)
```
执行效果如图7-4所示。

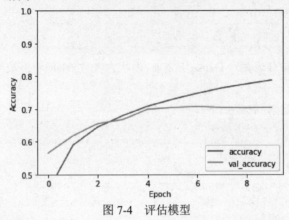

图 7-4　评估模型

然后通过如下代码显示评估结果：
```
print(test_acc)
```
执行后会输出：
```
0.7038999795913696
```

7.3.3　使用卷积神经网络进行图像分类

本节通过一个具体实例的实现过程，详细讲解使用卷积神经网络对花朵图像进行分类的过程。本实例使用 keras.Sequential 模型创建图像分类器，并使用 preprocessing.image_dataset_from_directory 加载数据。

实例 7-6：使用卷积神经网络对花朵图像进行分类

本实例将重点讲解如下两点：
- 加载并使用数据集；
- 识别过度拟合并应用技术来缓解它，包括数据增强和 Dropout。

1. 准备数据集

本实例的实现文件是 cnn02.py，使用大约 3 700 张鲜花照片的数据集，数据集包含 5 个子目录，每个类别一个目录：
```
flower_photo/
  daisy/
  dandelion/
  roses/
  sunflowers/
  tulips/
```

（1）下载数据集，代码如下：
```
import pathlib
dataset_url = "https://storage.googleapis.com/download.tensorflow.org/example_images/flower_photos.tgz"
data_dir = tf.keras.utils.get_file('flower_photos', origin=dataset_url, untar=True)
data_dir = pathlib.Path(data_dir)
image_count = len(list(data_dir.glob('*/*.jpg')))
```

```
print(image_count)
```
执行后会输出：
```
3670
```
这说明在数据集中共有 3 670 张图像。

（2）浏览数据集中 roses 目录中的第一个图像，代码如下：
```
roses = list(data_dir.glob('roses/*'))
PIL.Image.open(str(roses[0]))
```
执行后显示数据集中 roses 目录中的第一个图像，如图 7-5 所示。

（3）也可以浏览数据集中 tulips 目录中的第一个图像，代码如下：
```
tulips = list(data_dir.glob('tulips/*'))
PIL.Image.open(str(tulips[0]))
```
执行效果如图 7-6 所示。

图 7-5　roses 目录中的第一个图像　　　　图 7-6　tulips 目录中的第一个图像

2．创建数据集

使用 image_dataset_from_directory 从磁盘中加载数据集中的图像，然后从头开始编写自己的加载数据集代码。

（1）首先为加载器定义加载参数，代码如下：
```
batch_size = 32
img_height = 180
img_width = 180
```

（2）在现实中通常使用验证拆分法创建神经网络模型，本实例中将使用 80%的图像进行训练，使用 20%的图像进行验证。使用 80%的图像进行训练的代码如下：
```
train_ds = tf.keras.preprocessing.image_dataset_from_directory(
  data_dir,
  validation_split=0.2,
  subset="training",
  seed=123,
  image_size=(img_height, img_width),
  batch_size=batch_size)
```
执行后会输出：
```
Found 3670 files belonging to 5 classes.
Using 2936 files for training.
```
使用 20%的图像进行验证的代码如下：
```
val_ds = tf.keras.preprocessing.image_dataset_from_directory(
  data_dir,
  validation_split=0.2,
  subset="validation",
  seed=123,
```

```
    image_size=(img_height, img_width),
    batch_size=batch_size)
```

执行后会输出：

```
Found 3670 files belonging to 5 classes.
Using 734 files for validation.
```

可以在数据集的属性 class_names 中找到类名，每个类名和目录名称的字母顺序对应。例如下面的代码：

```
class_names = train_ds.class_names
print(class_names)
```

执行后会显示类名：

```
['daisy', 'dandelion', 'roses', 'sunflowers', 'tulips']
```

（3）可视化数据集中的数据，通过如下代码显示训练数据集中的前 9 张图像。

```
import matplotlib.pyplot as plt

plt.figure(figsize=(10, 10))
for images, labels in train_ds.take(1):
  for i in range(9):
    ax = plt.subplot(3, 3, i + 1)
    plt.imshow(images[i].numpy().astype("uint8"))
    plt.title(class_names[labels[i]])
    plt.axis("off")
```

执行效果如图 7-7 所示。

图 7-7　训练数据集中的前 9 张图像

（4）接下来将通过将这些数据集传递给训练模型 model.fit，也可以手动迭代数据集并检索批量图像。代码如下：

```
for image_batch, labels_batch in train_ds:
  print(image_batch.shape)
  print(labels_batch.shape)
  break
```

执行后会输出：

```
(32, 180, 180, 3)
(32,)
```

通过上述输出可知,image_batch 是形状的张量(32, 180, 180, 3)。这是一批 32 张形状图像:180×180×3(最后一个维度是指颜色通道 RGB),label_batch 是形状的张量(32,),这些都是对应标签 32 倍的图像。我们可以通过 numpy()在 image_batch 和 labels_batch 张量将上述图像转换为一个 numpy.ndarray。

3. 配置数据集

(1)接下来将配置数据集以提高性能,确保本实例使用缓冲技术以确保可以从磁盘生成数据,而不会导致 I/O 阻塞,下面是在加载数据时建议使用的两种重要方法:

- Dataset.cache():当从磁盘加载图像后,将图像保存在内存中。这将确保数据集在训练模型时不会成为瓶颈。如果数据集太大而无法放入内存,也可以使用此方法来创建高性能的磁盘缓存。
- Dataset.prefetch():在训练时重叠数据预处理和模型执行。

(2)进行数据标准化处理,因为 RGB 通道值在[0, 255]范围内,这对于神经网络来说并不理想。一般来说,应该设法使输入值变小。在本实例中将使用[0, 1]重新缩放图层将值标准化在范围内。

```
normalization_layer = layers.experimental.preprocessing.Rescaling(1./255)
```

(3)可以通过调用 map 将该层应用于数据集:

```
normalized_ds = train_ds.map(lambda x, y: (normalization_layer(x), y))
image_batch, labels_batch = next(iter(normalized_ds))
first_image = image_batch[0]
print(np.min(first_image), np.max(first_image))
```

执行后会输出:

```
0.0 0.9997713
```

或者,可以在模型定义中包含该层,这样可以简化部署,本实例将使用第二种方法。

4. 创建模型

本实例的模型由三个卷积块组成,每个块都有一个最大池层。有一个全连接层,上面有 128 个单元,由激活函数激活。该模型尚未针对高精度进行调整,本实例的目标是展示一种标准方法。代码如下:

```
num_classes = 5

model = Sequential([
   layers.experimental.preprocessing.Rescaling(1./255, input_shape=(img_height, img_width, 3)),
   layers.Conv2D(16, 3, padding='same', activation='relu'),
   layers.MaxPooling2D(),
   layers.Conv2D(32, 3, padding='same', activation='relu'),
   layers.MaxPooling2D(),
   layers.Conv2D(64, 3, padding='same', activation='relu'),
   layers.MaxPooling2D(),
   layers.Flatten(),
   layers.Dense(128, activation='relu'),
   layers.Dense(num_classes)
])
```

5. 编译模型

(1)在本实例中使用 optimizers.Adam 优化器和 losses.SparseCategoricalCrossentropy 损失函

数。要想查看每个训练时期的训练和验证准确性，需要传递 metrics 参数。代码如下：

```
model.compile(optimizer='adam',
              loss=tf.keras.losses.SparseCategoricalCrossentropy(from_logits=True),
              metrics=['accuracy'])
```

（2）使用模型的函数 summary 查看网络中的所有层，代码如下：

```
model.summary()
```

执行后会输出：

```
Model: "sequential"
_____
Layer (type)                 Output Shape              Param #
=================================================================
rescaling_1 (Rescaling)      (None, 180, 180, 3)       0

conv2d (Conv2D)              (None, 180, 180, 16)      448

max_pooling2d (MaxPooling2D) (None, 90, 90, 16)        0

conv2d_1 (Conv2D)            (None, 90, 90, 32)        4640

max_pooling2d_1 (MaxPooling2 (None, 45, 45, 32)        0

conv2d_2 (Conv2D)            (None, 45, 45, 64)        18496

max_pooling2d_2 (MaxPooling2 (None, 22, 22, 64)        0

flatten (Flatten)            (None, 30976)             0

dense (Dense)                (None, 128)               3965056

dense_1 (Dense)              (None, 5)                 645
=================================================================
Total params: 3,989,285
Trainable params: 3,989,285
Non-trainable params: 0
```

6. 训练模型

开始训练模型，代码如下：

```
epochs=10
history = model.fit(
  train_ds,
  validation_data=val_ds,
  epochs=epochs
)
```

执行后会输出：

```
Epoch 1/10
92/92 [==============================] - 3s 16ms/step - loss: 1.4412 - accuracy: 0.3784 - val_loss: 1.1290 - val_accuracy: 0.5409
Epoch 2/10
92/92 [==============================] - 1s 10ms/step - loss: 1.0614 - accuracy: 0.5841 - val_loss: 1.0058 - val_accuracy: 0.6131
Epoch 3/10
92/92 [==============================] - 1s 10ms/step - loss: 0.8999 - accuracy: 0.6560 - val_loss: 0.9920 - val_accuracy: 0.6104
Epoch 4/10
92/92 [==============================] - 1s 10ms/step - loss: 0.7416 - accuracy: 0.7153 - val_loss: 0.9279 - val_accuracy: 0.6458
Epoch 5/10
```

```
    92/92 [==============================] - 1s 10ms/step - loss: 0.5618 - accuracy:
0.7844 - val_loss: 1.0019 - val_accuracy: 0.6322
    Epoch 6/10
    92/92 [==============================] - 1s 10ms/step - loss: 0.3950 - accuracy:
0.8634 - val_loss: 1.0232 - val_accuracy: 0.6553
    Epoch 7/10
    92/92 [==============================] - 1s 10ms/step - loss: 0.2228 - accuracy:
0.9268 - val_loss: 1.2722 - val_accuracy: 0.6444
    Epoch 8/10
    92/92 [==============================] - 1s 10ms/step - loss: 0.1188 - accuracy:
0.9687 - val_loss: 1.4410 - val_accuracy: 0.6567
    Epoch 9/10
    92/92 [==============================] - 1s 10ms/step - loss: 0.0737 - accuracy:
0.9802 - val_loss: 1.6363 - val_accuracy: 0.6444
    Epoch 10/10
    92/92 [==============================] - 1s 10ms/step - loss: 0.0566 - accuracy:
0.9847 -
```

7．可视化训练结果

在训练集和验证集上创建损失图和准确度图，然后绘制可视化结果，代码如下：

```
acc = history.history['accuracy']
val_acc = history.history['val_accuracy']

loss = history.history['loss']
val_loss = history.history['val_loss']

epochs_range = range(epochs)

plt.figure(figsize=(8, 8))
plt.subplot(1, 2, 1)
plt.plot(epochs_range, acc, label='Training Accuracy')
plt.plot(epochs_range, val_acc, label='Validation Accuracy')
plt.legend(loc='lower right')
plt.title('Training and Validation Accuracy')

plt.subplot(1, 2, 2)
plt.plot(epochs_range, loss, label='Training Loss')
plt.plot(epochs_range, val_loss, label='Validation Loss')
plt.legend(loc='upper right')
plt.title('Training and Validation Loss')
plt.show()
```

执行后的效果如图7-8所示。

8．过拟合处理：数据增强

从可视化损失图和准确度图中的执行效果可以看出，训练准确率和验证准确率相差很大，模型在验证集上的准确率只有60%左右。训练准确度随着时间线性增加，而验证准确度在训练过程中停滞在60%左右。此外，训练和验证准确性之间的准确性差异是显而易见的，这是过度拟合的迹象。

当训练样例数量较少时，模型有时会从训练样例中的噪声或不需要的细节中学习，这在一定程度上会对模型在新样例上的性能产生负面影响。这种现象被称为过拟合。这意味着该模型将很难在新数据集上泛化。在训练过程中有多种方法可以对抗过度拟合。

过拟合通常发生在训练样本较少时，数据增强采用的方法是从现有示例中生成额外的训练数据，方法是使用随机变换来增强它们，从而产生看起来可信的图像。这有助于将模型暴露于数据的更多方面并更好地概括。

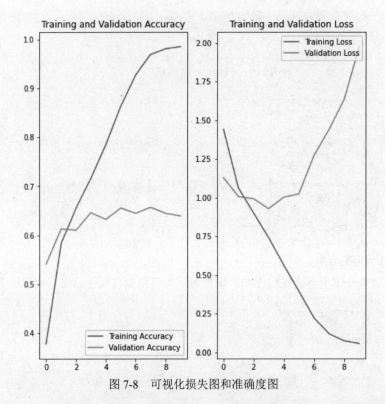

图 7-8 可视化损失图和准确度图

通过使用 tf.keras.layers.experimental.preprocessing 实现数据增强，可以像其他层一样包含在模型中，并在 GPU 上运行。代码如下：

```
data_augmentation = keras.Sequential(
  [
    layers.experimental.preprocessing.RandomFlip("horizontal",
input_shape=(img_height, img_width,3)),
    layers.experimental.preprocessing.RandomRotation(0.1),
    layers.experimental.preprocessing.RandomZoom(0.1),
  ]
)
```

此时可以对同一图像多次应用数据增强技术。下面是可视化数据增强的代码：

```
plt.figure(figsize=(10, 10))
for images, _ in train_ds.take(1):
  for i in range(9):
    augmented_images = data_augmentation(images)
    ax = plt.subplot(3, 3, i + 1)
    plt.imshow(augmented_images[0].numpy().astype("uint8"))
    plt.axis("off")
```

执行后的效果如图 7-9 所示。

9. 过拟合处理：将 Dropout 引入网络

接下来介绍另一种减少过拟合的技术——将 Dropout 引入网络。这是一种正则化处理形式。当将 Dropout 应用于一个层时，它会在训练过程中从该层中随机删除（通过将激活设置为零）许多输出单元。Dropout 将一个小数作为其输入值，例如 0.1、0.2、0.4 等，这意味着从应用层中随机丢弃 10%、20% 或 40% 的输出单元。下面的代码是创建一个新的神经网络 layers.Dropout，然后使用增强图像对其进行训练。

```
model = Sequential([
  data_augmentation,
```

```
  layers.experimental.preprocessing.Rescaling(1./255),
  layers.Conv2D(16, 3, padding='same', activation='relu'),
  layers.MaxPooling2D(),
  layers.Conv2D(32, 3, padding='same', activation='relu'),
  layers.MaxPooling2D(),
  layers.Conv2D(64, 3, padding='same', activation='relu'),
  layers.MaxPooling2D(),
  layers.Dropout(0.2),
  layers.Flatten(),
  layers.Dense(128, activation='relu'),
  layers.Dense(num_classes)
])
```

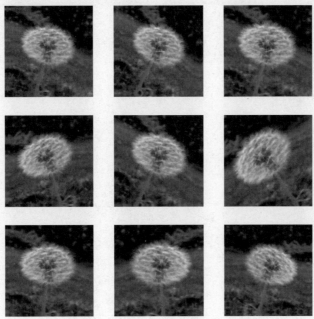

图 7-9　数据增强

10．重新编译和训练模型

经过前面的过拟合处理，接下来重新编译和训练模型，重新编译模型的代码如下：

```
model.compile(optimizer='adam',
              loss=tf.keras.losses.SparseCategoricalCrossentropy(from_logits=True),
              metrics=['accuracy'])
model.summary()
Model: "sequential_2"
```

执行后会输出：

```
Layer (type)                 Output Shape              Param #
=================================================================
sequential_1 (Sequential)    (None, 180, 180, 3)       0
_____
rescaling_2 (Rescaling)      (None, 180, 180, 3)       0
_____
conv2d_3 (Conv2D)            (None, 180, 180, 16)      448
_____
max_pooling2d_3 (MaxPooling2 (None, 90, 90, 16)        0
_____
conv2d_4 (Conv2D)            (None, 90, 90, 32)        4640
_____
max_pooling2d_4 (MaxPooling2 (None, 45, 45, 32)        0
```

```
conv2d_5 (Conv2D)            (None, 45, 45, 64)         18496

max_pooling2d_5 (MaxPooling2 (None, 22, 22, 64)         0

dropout (Dropout)            (None, 22, 22, 64)         0

flatten_1 (Flatten)          (None, 30976)              0

dense_2 (Dense)              (None, 128)                3965056

dense_3 (Dense)              (None, 5)                  645
=================================================================
Total params: 3,989,285
Trainable params: 3,989,285
Non-trainable params: 0
```

重新训练模型的代码如下：

```
epochs = 15
history = model.fit(
  train_ds,
  validation_data=val_ds,
  epochs=epochs
)
```

执行后会输出：

```
Epoch 1/15
92/92 [==============================] - 2s 13ms/step - loss: 1.2685 - accuracy: 0.4465 - val_loss: 1.0464 - val_accuracy: 0.5899
Epoch 2/15
92/92 [==============================] - 1s 11ms/step - loss: 1.0195 - accuracy: 0.5964 - val_loss: 0.9466 - val_accuracy: 0.6008
Epoch 3/15
92/92 [==============================] - 1s 11ms/step - loss: 0.9184 - accuracy: 0.6356 - val_loss: 0.8412 - val_accuracy: 0.6689
Epoch 4/15
92/92 [==============================] - 1s 11ms/step - loss: 0.8497 - accuracy: 0.6768 - val_loss: 0.9339 - val_accuracy: 0.6444
Epoch 5/15
92/92 [==============================] - 1s 11ms/step - loss: 0.8180 - accuracy: 0.6781 - val_loss: 0.8309 - val_accuracy: 0.6689
Epoch 6/15
92/92 [==============================] - 1s 11ms/step - loss: 0.7424 - accuracy: 0.7105 - val_loss: 0.7765 - val_accuracy: 0.6962
Epoch 7/15
92/92 [==============================] - 1s 11ms/step - loss: 0.7157 - accuracy: 0.7251 - val_loss: 0.7451 - val_accuracy: 0.7016
Epoch 8/15
92/92 [==============================] - 1s 11ms/step - loss: 0.6764 - accuracy: 0.7476 - val_loss: 0.9703 - val_accuracy: 0.6485
Epoch 9/15
92/92 [==============================] - 1s 11ms/step - loss: 0.6667 - accuracy: 0.7439 - val_loss: 0.7249 - val_accuracy: 0.6962
Epoch 10/15
92/92 [==============================] - 1s 11ms/step - loss: 0.6282 - accuracy: 0.7619 - val_loss: 0.7187 - val_accuracy: 0.7071
Epoch 11/15
92/92 [==============================] - 1s 11ms/step - loss: 0.5816 - accuracy: 0.7793 - val_loss: 0.7107 - val_accuracy: 0.7275
Epoch 12/15
92/92 [==============================] - 1s 11ms/step - loss: 0.5570 - accuracy: 0.7813 - val_loss: 0.6945 - val_accuracy: 0.7493
Epoch 13/15
```

```
       92/92 [==============================] - 1s 11ms/step - loss: 0.5396 - accuracy:
0.7939 - val_loss: 0.6713 - val_accuracy: 0.7302
    Epoch 14/15
       92/92 [==============================] - 1s 11ms/step - loss: 0.5194 - accuracy:
0.7936 - val_loss: 0.6771 - val_accuracy: 0.7371
    Epoch 15/15
       92/92 [==============================] - 1s 11ms/step - loss: 0.4930 - accuracy:
0.8096 - val_loss: 0.6705 - val_accuracy: 0.7384
```

在使用数据增强和 Dropout 处理后，过拟合比以前少了，训练和验证的准确性更接近。接下来重新可视化训练结果，代码如下：

```
acc = history.history['accuracy']
val_acc = history.history['val_accuracy']

loss = history.history['loss']
val_loss = history.history['val_loss']

epochs_range = range(epochs)

plt.figure(figsize=(8, 8))
plt.subplot(1, 2, 1)
plt.plot(epochs_range, acc, label='Training Accuracy')
plt.plot(epochs_range, val_acc, label='Validation Accuracy')
plt.legend(loc='lower right')
plt.title('Training and Validation Accuracy')

plt.subplot(1, 2, 2)
plt.plot(epochs_range, loss, label='Training Loss')
plt.plot(epochs_range, val_loss, label='Validation Loss')
plt.legend(loc='upper right')
plt.title('Training and Validation Loss')
plt.show()
```

代码执行后的效果如图 7-10 所示。

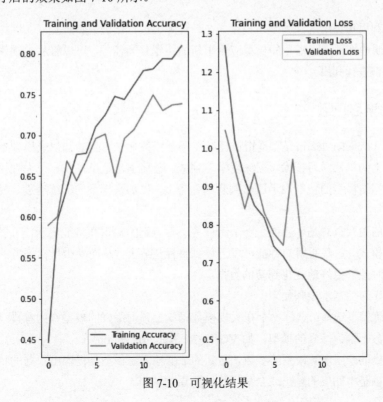

图 7-10　可视化结果

11．预测新数据

最后使用最新创建的模型对未包含在训练或验证集中的图像进行分类处理，代码如下：

```
sunflower_url = "https://storage.googleapis.com/download.tensorflow.org/example_images/ 592px-Red_sunflower.jpg"
sunflower_path = tf.keras.utils.get_file('Red_sunflower', origin=sunflower_url)

img = keras.preprocessing.image.load_img(
    sunflower_path, target_size=(img_height, img_width)
)
img_array = keras.preprocessing.image.img_to_array(img)
img_array = tf.expand_dims(img_array, 0) # Create a batch

predictions = model.predict(img_array)
score = tf.nn.softmax(predictions[0])

print(
    "This image most likely belongs to {} with a {:.2f} percent confidence."
    .format(class_names[np.argmax(score)], 100 * np.max(score))
)
```

执行后会输出：

```
Downloading data from https://storage.googleapis.com/download.tensorflow.org/example_images/592px-Red_sunflower.jpg
122880/117948 [==============================] - 0s 0us/step
This image most likely belongs to sunflowers with a 99.36 percent confidence.
```

大家需要注意的是，数据增强和Dropout层在推理时处于非活动状态。

7.4 基于迁移学习的图像分类

迁移学习利用预训练的深度学习模型，在新任务上进行微调或特征提取，以加快训练过程并提高性能。通过将预训练模型的权重应用于新的图像分类任务，可以将从大规模数据集中学习到的特征知识进行利用。

7.4.1 迁移学习介绍

迁移学习（transfer learning）是指将已经在一个任务上学习到的知识或模型参数应用到另一个相关任务上的机器学习技术。在深度学习领域，迁移学习是指将在一个大规模数据集上预训练好的神经网络模型的特征表示迁移到新的任务上，以加快模型的训练速度、提高泛化能力和提升性能。

迁移学习的主要思想是，通过在一个源任务上学习到的知识和特征表示，可以提取到数据的一般特征和模式。这些通用的特征可以迁移到目标任务上，从而减少目标任务的训练时间和数据需求，同时能够更好地泛化到新的数据。

实现迁移学习的基本步骤如下：

（1）预训练模型选择：选择一个在大规模数据集上预训练好的模型作为源模型。通常选择在图像分类任务上预训练好的模型，如VGG、ResNet、Inception等。

（2）特征提取：固定源模型的参数，将新的数据输入源模型，提取出数据的特征表示。这些特征表示可以是中间层的输出或全局平均池化层的输出。

(3)新模型构建:在源模型的基础上构建新的模型,通常是在提取的特征表示之上添加一个或多个全连接层进行分类。

(4)微调(fine-tuning):根据目标任务的数据,对新模型进行训练。可以选择解冻源模型的一部分或全部参数,并联合训练源模型和新模型。

通过迁移学习,可以充分利用源任务上学习到的知识和模型参数,加快模型的训练速度,提高模型的性能和泛化能力。特别是在数据集较小的情况下,迁移学习可以是解决过拟合和数据稀缺问题的一个有效的方法。

7.4.2 基于迁移学习的图片分类器

本节的实例是使用 TensorFlow.js 在浏览器中实现一个实时训练的分类器。首先,加载并运行一个名为 MobileNet 的常用预训练模型,用于在浏览器中进行图片分类。然后,使用"迁移学习"技术,该技术使用预训练的 MobileNet 模型对训练进行 Bootstrap 处理,并自定义该模型以对应用进行训练。

实例 7-7:使用 TensorFlow.js 实现一个实时训练的分类器

通过本实例可以掌握如下知识点:
- 如何加载预训练的 MobileNet 模型并利用新数据进行预测;
- 如何通过网络摄像头进行预测;
- 如何使用 MobileNet 的中间激活功能,使用网络摄像头即时为定义的一组新类别执行迁移学习。

1. 加载 TensorFlow. js 和 MobileNet 模型

编写 HTML 文件 index.html,在文件中使用 TensorFlow.js 和 MobileNet 模型,代码如下:

```html
<html>
  <head>
    <!-- Load the latest version of TensorFlow.js -->
    <script src="https://cdn.jsdelivr.net/npm/@tensorflow/tfjs"></script>
    <script src="https://cdn.jsdelivr.net/npm/@tensorflow-models/mobilenet"></script>
  </head>
  <body>
    <div id="console"></div>
    <!-- Add an image that we will use to test -->
    <img id="img" crossorigin src="https://i.imgur.com/JlUvsxa.jpg" width="227" height="227"/>
    <!-- Load index.js after the content of the page -->
    <script src="index.js"></script>
  </body>
</html>
```

2. 设置 MobileNet

编写文件 index.js,设置 MobileNet 以在浏览器中进行判断识别。代码如下:

```
let net;

async function app() {
  console.log('Loading mobilenet..');
```

```
  //加载模型
  net = await mobilenet.load();
  console.log('Successfully loaded model');

  //通过模型对图像进行预测
  const imgEl = document.getElementById('img');
  const result = await net.classify(imgEl);
  console.log(result);
}

app();
```

在浏览器中运行文件 index.html，会在开发者工具的 JavaScript 控制台中看到一张狗狗的图片，这是 MobileNet 预测的最有可能的内容。注意，下载模型可能需要一些时间，请耐心等待。另外，此模型也可以在手机上使用。

3．识别摄像头图片

接下来通过摄像头图片在浏览器中运行 MobileNet 判断识别。设置摄像头，对摄像头拍摄的图片进行预测。首先，设置网络摄像头视频元素。

（1）打开文件 index.html，将以下代码行添加到<body>部分内，然后删除用于加载狗狗图片的<img>标记：

```
<video autoplay playsinline muted id="webcam" width="224" height="224"></video>
```

（2）打开文件 index.js，将 camwareElement 添加到文件的最顶部。代码如下：

```
const webcamElement = document.getElementById('webcam');
```

现在，在之前添加的函数 app()中，移除通过图片获得的预测结果，在无限循环中创建一个与摄像头连接的元素并进行预测。函数 app()的新代码如下：

```
async function app() {
  console.log('Loading mobilenet..');

  //加载模型
  net = await mobilenet.load();
  console.log('Successfully loaded model');

  //从 Tensorflow.js 数据 API 创建一个对象，该对象可以作为 Tensor 从网络摄像机捕获图像
  const webcam = await tf.data.webcam(webcamElement);
  while (true) {
    const img = await webcam.capture();
    const result = await net.classify(img);

    document.getElementById('console').innerText = `
      prediction: ${result[0].className}\n
      probability: ${result[0].probability}
    `;
    //处理张量以释放内存
    img.dispose();

    //等待下一个动画帧触发
    await tf.nextFrame();
  }
}
```

如果此时在浏览器中打开控制台，会看到关于摄像头采集到的每个帧是什么的概率的 MobileNet 预测。这些预测可能是无意义的，因为 ImageNet 数据集中的内容与网络摄像头通常会捕捉的图片并不相似。要进行测试，可以在手机上显示一张图片，然后放在笔记本电脑摄像头前方。

4. 自定义分类器

接下来在 MobileNet 预测的基础上添加一个自定义分类器,制作一个即时使用网络摄像头的自定义 3 个类别的对象分类器。依旧通过 MobileNet 进行分类,但这次将对特定摄像头图片的模型进行内部表示(激活),并用它进行分类。使用一个名为"KNN 分类器"的模块,它能有效地将网络摄像头图片(实际上是它们的 MobileNet 激活)归到不同的类别。当用户要求进行预测时,只需要选择拥有要为它进行预测的图片最相似的激活的类别。

(1)在文件 index.html 中的<head>标记中的末尾添加 KNN 分类器的导入项(仍需要 MobileNet,因此请勿移除该导入项):

```
<script src="https://cdn.jsdelivr.net/npm/@tensorflow-models/knn-classifier"></script>
```

然后为在文件 index.html 中的<video>标记下面添加 3 个按钮,这些按钮将用于向模型中添加训练图片。代码如下:

```
...
<button id="class-a">Add A</button>
<button id="class-b">Add B</button>
<button id="class-c">Add C</button>
...
```

(2)在文件 index.js 的顶部创建分类器,代码如下:

```
const classifier = knnClassifier.create();
```

然后修改函数 app(),代码如下:

```
async function app() {
  console.log('Loading mobilenet..');

  //加载模型
  net = await mobilenet.load();
  console.log('Successfully loaded model');

  //从Tensorflow.js API 创建一个对象,该对象可以作为 Tensor 从摄像机捕获图像
  const webcam = await tf.data.webcam(webcamElement);

  //从摄像头读取图像并将它与特定的类索引相关联
  const addExample = async classId => {
    //从摄像头捕获图像
    const img = await webcam.capture();

    //获取 MobileNet'conv\u preds'的中间激活,并将它传递给 KNN 分类器
    const activation = net.infer(img, true);

    //将中间激活传递给分类器
    classifier.addExample(activation, classId);

    //释放张量.
    img.dispose();
  };

  //单击按钮时,为该类添加一个示例
  document.getElementById('class-a').addEventListener('click', () => addExample(0));
  document.getElementById('class-b').addEventListener('click', () => addExample(1));
  document.getElementById('class-c').addEventListener('click', () => addExample(2));

  while (true) {
```

```javascript
    if (classifier.getNumClasses() > 0) {
      const img = await webcam.capture();

      //从摄像头获取mobilenet的激活
      const activation = net.infer(img, 'conv_preds');
      //从分类器模块中获取最可能的类和可信度
      const result = await classifier.predictClass(activation);

      const classes = ['A', 'B', 'C'];
      document.getElementById('console').innerText = `
        prediction: ${classes[result.label]}\n
        probability: ${result.confidences[result.label]}
      `;

      //释放张量
      img.dispose();
    }

    await tf.nextFrame();
  }
}
```

现在，当在浏览器中运行文件 index.html，可以使用常见物体或"面部/身体"手势为这三个类别捕获图片。当每次单击某一个"添加"按钮时，都会将一张图片作为一个示例添加到该类别中。在执行此操作时，模型会针对传入的摄像头图片持续进行预测，并实时显示预测结果。

7.5 基于循环神经网络的图像分类

循环神经网络（recurrent neural network，RNN）主要应用于序列数据的处理，但也可以用于图像分类中。例如，可以将图像视为像素序列，然后使用 RNN 进行序列建模和分类。

7.5.1 循环神经网络介绍

循环神经网络是一种用于处理序列数据的神经网络模型。与传统的前馈神经网络不同，RNN 具有循环结构，可以保留和利用序列数据中的时间信息。

在传统的前馈神经网络中，每个输入和输出之间是独立的，而 RNN 引入了一个隐藏状态（hidden state）来保存过去的信息，并在当前时间步骤使用它。RNN 的隐藏状态可以看作是网络的记忆，它可以将过去的信息传递到当前时间步骤，并对当前的输入进行处理。

RNN 的一个关键特点是它能够处理任意长度的序列输入。它通过将权重共享的方式，将相同的网络结构应用于不同的时间步骤上。这样，RNN 可以在处理每个时间步骤的输入时，共享参数并保留序列数据的上下文信息。

RNN 的基本结构是一个循环单元（recurrent unit），通常使用 tanh 或 ReLU 等激活函数来处理输入和隐藏状态。常见的 RNN 变体包括长短期记忆网络（long short-term memory，简称 LSTM）和门控循环单元（gated recurrent unit，GRU），它们在处理长期依赖性和梯度消失问题上更加有效。

RNN 广泛应用于自然语言处理（例如语言模型、机器翻译和文本生成）、语音识别、时间序列预测、图像描述生成等任务，其中序列数据的时间依赖性是关键。

7.5.2 实战演练

下面是一个使用循环神经网络实现图像分类的例子，其中使用了 LSTM 作为 RNN 的基本单元。

实例7-8：使用循环神经网络实现图像分类

源码路径：daima\7\xun.py

```python
import numpy as np
import tensorflow as tf
from tensorflow import keras

# 加载并准备数据集
(train_images, train_labels), (test_images, test_labels) = keras.datasets.mnist.load_data()
train_images = train_images / 255.0
test_images = test_images / 255.0

# 将图像转换为序列数据
train_sequences = train_images.reshape(train_images.shape[0], -1, 28)  # 将图像展平并将每个图像看作是一个序列
test_sequences = test_images.reshape(test_images.shape[0], -1, 28)

# 定义循环神经网络模型
model = keras.Sequential([
    keras.layers.LSTM(64, input_shape=(None, 28)),  # LSTM层用于处理序列数据
    keras.layers.Dense(10, activation='softmax')  # 输出层，对应类别数目
])

# 编译和训练模型
model.compile(optimizer='adam', loss='sparse_categorical_crossentropy', metrics=['accuracy'])
model.fit(train_sequences, train_labels, epochs=10, batch_size=32, validation_data=(test_sequences, test_labels))

# 评估模型
test_loss, test_acc = model.evaluate(test_sequences, test_labels)
print('Test accuracy:', test_acc)
```

在上述代码中，使用 MNIST 数据集作为示例数据集。首先，将图像展平为序列，即将每个图像的每一行作为序列中的一个时间步骤。然后，定义了一个包含 LSTM 层和输出层的循环神经网络模型。使用 Adam 优化器和稀疏交叉熵损失函数编译模型，并在训练数据上训练模型。最后，使用测试数据评估模型的准确率。

注意：实际上，在图像分类任务中，CNN 是更常用和有效的选择，因为 CNN 能够更好地捕捉图像的空间特征。RNN 主要用于处理序列数据，例如自然语言处理任务或时间序列预测任务。但是，如果图像中存在时间依赖关系或需要处理图像序列数据，那么可以尝试使用 RNN 进行图像分类。

7.6 基于卷积循环神经网络的图像分类

RNN 在图像处理任务中的应用相对较少，因为传统的 RNN 架构不太适用于处理图像数据。

然而，有一些扩展和变种的 RNN 模型可以用于图像处理任务，特别是在处理与时间有关的图像序列时。一种常见的 RNN 变体是卷积循环神经网络（convolutional recurrent neural network，CRNN）。

7.6.1 卷积循环神经网络介绍

CRNN 结合了 CNN 和 RNN 的特性，主要用于处理图像序列数据，例如图像标注、场景文本识别等任务。

CRNN 的主要思想是将 CNN 用于提取图像的空间特征，然后将提取的特征序列输入到 RNN 中，以捕捉图像序列的上下文信息。通过这种方式，CRNN 能够同时考虑到图像的空间结构和时间依赖关系。CRNN 的整体结构通常包含以下几个关键组件：

- 卷积层（convolutional layer）：使用卷积操作提取图像的空间特征。卷积层通常包括多个卷积核（filters）以捕捉不同的特征。在图像序列任务中，卷积核的宽度通常等于图像的宽度，而高度则可以根据需要调整。
- 池化层（pooling layer）：用于降低特征图的维度，并提取最显著的特征。池化操作通常通过取局部区域内的最大值或平均值来实现。
- 循环层（recurrent layer）：通常使用循环神经网络（如 LSTM 或 GRU）来处理卷积层输出的特征序列。循环层能够捕捉到序列数据的时间依赖关系，并生成具有上下文信息的隐藏表示。
- 全连接层（fully connected layer）：用于将循环层输出的隐藏表示映射到类别标签。全连接层可以包括多个神经元，并使用适当的激活函数（如 softmax）来输出分类概率。

CRNN 的训练过程通常是端到端的，通过反向传播算法更新网络参数。在训练之前，需要准备好带有标签的图像序列数据集，并对图像进行预处理（如归一化、调整大小等）。

CRNN 在图像序列任务中具有广泛的应用，如场景文本识别、图像标注、视频描述生成等。通过结合 CNN 和 RNN 的特性，CRNN 能够充分利用图像序列数据中的空间和时间信息，从而在处理图像序列任务时取得良好的效果。

7.6.2 CRNN 图像识别器

请看下面的实例，功能是使用卷积循环神经网络模型（CRNN）对 CIFAR-10 数据集进行图像识别。在本实例中定义了一个卷积循环神经网络模型（CRNN），该模型由卷积层、池化层、LSTM 层和全连接层组成。代码加载并预处理 CIFAR-10 数据集，使用交叉熵损失函数和 Adam 优化器进行模型训练，然后在测试集上评估模型的准确性。

实例 7-9：使用卷积循环神经网络模型对数据集进行图像识别

源码路径： daima\7\shipin.py

```
import torch
import torch.nn as nn
import torch.optim as optim
import torchvision
import torchvision.transforms as transforms
```

```python
# 定义卷积循环神经网络模型
class CRNN(nn.Module):
    def __init__(self):
        super(CRNN, self).__init__()
        self.conv1 = nn.Conv2d(3, 64, kernel_size=3, stride=1, padding=1)
        self.relu = nn.ReLU()
        self.pool = nn.MaxPool2d(kernel_size=2, stride=2)
        self.lstm = nn.LSTM(64 * 8 * 8, 128, batch_first=True)
        self.fc = nn.Linear(128, 10)

    def forward(self, x):
        x = self.relu(self.conv1(x))
        x = self.pool(x)
        x = x.view(x.size(0), -1)
        x = x.unsqueeze(1)
        x, _ = self.lstm(x)
        x = x[:, -1, :]
        x = self.fc(x)
        return x

# 设置训练参数
batch_size = 32
lr = 0.001
num_epochs = 10

# 加载和预处理数据
transform = transforms.Compose([
    transforms.ToTensor(),
    transforms.Normalize((0.5, 0.5, 0.5), (0.5, 0.5, 0.5))
])

train_dataset = torchvision.datasets.CIFAR10(root='./data', train=True, download=True, transform=transform)
train_loader = torch.utils.data.DataLoader(train_dataset, batch_size=batch_size, shuffle=True)

test_dataset = torchvision.datasets.CIFAR10(root='./data', train=False, download=True, transform=transform)
test_loader = torch.utils.data.DataLoader(test_dataset, batch_size=batch_size, shuffle=False)

# 创建模型实例
model = CRNN()

# 定义损失函数和优化器
criterion = nn.CrossEntropyLoss()
optimizer = optim.Adam(model.parameters(), lr=lr)

# 训练模型
total_step = len(train_loader)
for epoch in range(num_epochs):
    for i, (images, labels) in enumerate(train_loader):
        # 前向传播
        outputs = model(images)
        loss = criterion(outputs, labels)
```

```
    # 反向传播和优化
    optimizer.zero_grad()
    loss.backward()
    optimizer.step()

    # 每100个批次打印一次训练状态
    if (i+1) % 100 == 0:
        print(f'Epoch [{epoch+1}/{num_epochs}], Step [{i+1}/{total_step}], Loss: {loss.item():.4f}')

# 测试模型
model.eval()
with torch.no_grad():
    correct = 0
    total = 0
    for images, labels in test_loader:
        outputs = model(images)
        _, predicted = torch.max(outputs.data, 1)
        total += labels.size(0)
        correct += (predicted == labels).sum().item()

    accuracy = 100 * correct / total
    print(f'Test Accuracy: {accuracy:.2f} %')
```

上述代码的实现流程如下：

（1）导入所需的 PyTorch 库：torch 用于核心功能；torch.nn 用于定义神经网络模型；torch.optim 用于定义优化器；torchvision 用于数据集和数据转换操作。

（2）定义 CRNN 模型：创建一个继承自 nn.Module 的类 CRNN，其中定义了卷积层、激活函数、池化层、LSTM 层和全连接层。这些层构成了 CRNN 模型的结构。

（3）设置训练参数：定义批量大小（batch_size）、学习率（lr）和训练轮数（num_epochs）。

（4）加载和预处理数据：使用 torchvision 库加载 CIFAR-10 数据集，并定义数据预处理操作。数据集被分为训练集和测试集，并使用 DataLoader 将其转换为可迭代的数据加载器。

（5）创建模型实例：实例化 CRNN 类，创建 CRNN 模型。

（6）定义损失函数和优化器：使用交叉熵损失函数 nn.CrossEntropyLoss()和 Adam 优化器 optim.Adam()来定义训练过程中使用的损失函数和优化算法。

（7）训练模型：通过迭代训练数据加载器中的批次数据，进行前向传播、计算损失、反向传播和优化权重的过程。每 100 个批次打印一次训练状态。

（8）测试模型：将模型设置为评估模式 model.eval()，并通过迭代测试数据加载器中的批次数据，进行前向传播并计算预测精度。

上述代码展示了使用卷积循环神经网络进行图像分类的基本流程，包括模型定义、数据加载和预处理、训练和测试过程。通过调整参数和网络结构，可以应用于其他图像分类任务。

第 8 章 鲜花识别系统开发

本章通过一个鲜花识别系统的实现过程，详细讲解使用 TensorFlow Lite 开发大型软件项目的过程，包括项目的架构分析、创建模型和具体实现知识。

8.1 系统介绍

机器学习已成为移动开发中的重要工具，为现代移动应用程序提供了许多智能功能。本项目是基于 Codelab 开发机器学习模型，该模型可以使用 TensorFlow 识别鲜花图像，然后将这个模型部署到 Android 应用程序。在手机中通过摄像头采集鲜花照片，可以实时识别鲜花的名字。本项目的具体结构如图 8-1 所示。

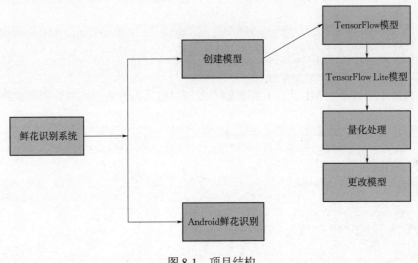

图 8-1 项目结构

8.2 创建模型

在创建鲜花识别系统之前，需要先创建识别模型。先使用 TensorFlow 创建普通的数据模型，然后转换为 TensorFlow Lite 数据模型。本项通过文件 mo.py 创建模型，下面详细讲解这个模型文件的具体实现过程。

8.2.1 创建 TensorFlow 数据模型

（1）首先安装 GitHub repo 中的 Model Maker 库，具体方法如下：

```
pip install -q tflite-model-maker
```

然后导入本项目需要的库：

```python
import os

import numpy as np

import tensorflow as tf
assert tf.__version__.startswith('2')

from tflite_model_maker import model_spec
from tflite_model_maker import image_classifier
from tflite_model_maker.config import ExportFormat
from tflite_model_maker.config import QuantizationConfig
from tflite_model_maker.image_classifier import DataLoader

import matplotlib.pyplot as plt
```

（2）获取数据路径，代码如下：

```python
image_path = tf.keras.utils.get_file(
      'flower_photos.tgz',
      'https://storage.googleapis.com/download.tensorflow.org/example_images/flower_photos.tgz',
      extract=True)
image_path = os.path.join(os.path.dirname(image_path), 'flower_photos')
```

执行后会输出：

```
Downloading data from https://storage.googleapis.com/download.tensorflow.org/example_images/flower_photos.tgz
228818944/228813984 [==============================] - 63s 0us/step
228827136/228813984 [==============================] - 63s 0us/step
```

（3）加载特定于设备上 ML 应用程序的输入数据，将其拆分为训练数据和测试数据。代码如下：

```python
data = DataLoader.from_folder(image_path)
train_data, test_data = data.split(0.9)
```

执行后会输出：

```
2023-09-12 11:22:56.386698: I tensorflow/stream_executor/cuda/cuda_gpu_executor.cc:937] successful NUMA node read from SysFS had negative value (-1), but there must be at least one NUMA node, so returning NUMA node zero
INFO:tensorflow:Load image with size: 3670, num_label: 5, labels: daisy, dandelion, roses, sunflowers, tulips.
2023-09-12 11:22:56.395523: I tensorflow/stream_executor/cuda/cuda_gpu_executor.cc:937] successful NUMA node read from SysFS had negative value (-1), but there must be at least one NUMA node, so returning NUMA node zero
2023-09-12 11:22:56.396549: I tensorflow/stream_executor/cuda/cuda_gpu_executor.cc:937] successful NUMA node read from SysFS had negative value (-1), but there must be at least one NUMA node, so returning NUMA node zero
2023-09-12 11:22:56.398220: I tensorflow/core/platform/cpu_feature_guard.cc:142] This TensorFlow binary is optimized with oneAPI Deep Neural Network Library (oneDNN) to use the following CPU instructions in performance-critical operations:  AVX2 AVX512F FMA
To enable them in other operations, rebuild TensorFlow with the appropriate compiler flags.
2023-09-12 11:22:56.398875: I tensorflow/stream_executor/cuda/cuda_gpu_executor.cc:937] successful NUMA node read from SysFS had negative value (-1), but there must be at least one NUMA node, so returning NUMA node zero
2023-09-12 11:22:56.400004: I tensorflow/stream_executor/cuda/cuda_gpu_executor.cc:937] successful NUMA node read from SysFS had negative value (-1), but there must be at least one NUMA node, so returning NUMA node zero
```

```
2023-09-12 11:22:56.400967: I tensorflow/stream_executor/cuda/cuda_gpu_executor.
cc:937] successful NUMA node read from SysFS had negative value (-1), but there must be
at least one NUMA node, so returning NUMA node zero
2023-09-12 11:22:57.007249: I tensorflow/stream_executor/cuda/cuda_gpu_executor.
cc:937] successful NUMA node read from SysFS had negative value (-1), but there must be
at least one NUMA node, so returning NUMA node zero
2023-09-12 11:22:57.008317: I tensorflow/stream_executor/cuda/cuda_gpu_executor.
cc:937] successful NUMA node read from SysFS had negative value (-1), but there must be
at least one NUMA node, so returning NUMA node zero
2023-09-12 11:22:57.009214: I tensorflow/stream_executor/cuda/cuda_gpu_executor.
cc:937] successful NUMA node read from SysFS had negative value (-1), but there must be
at least one NUMA node, so returning NUMA node zero
2023-09-12 11:22:57.010137: I tensorflow/core/common_runtime/gpu/gpu_device.cc:
1510] Created device /job:localhost/replica:0/task:0/device:GPU:0 with 14648 MB memory:
-> device: 0, name: Tesla V100-SXM2-16GB, pci bus id: 0000:00:05.0, compute capability: 7.0
```

(4) 自定义 TensorFlow 模型，代码如下：

```
model = image_classifier.create(train_data)
```

执行后会输出：

```
INFO:tensorflow:Retraining the models...
2023-09-12 11:23:00.961952: I tensorflow/compiler/mlir/mlir_graph_optimization_
pass.cc:185] None of the MLIR Optimization Passes are enabled (registered 2)
Model: "sequential"
_____
 Layer (type)                Output Shape              Param #
=================================================================
 hub_keras_layer_v1v2 (HubKer (None, 1280)             3413024

 dropout (Dropout)           (None, 1280)              0

 dense (Dense)               (None, 5)                 6405
=================================================================
Total params: 3,419,429
Trainable params: 6,405
Non-trainable params: 3,413,024
_____

None
Epoch 1/5
/tmpfs/src/tf_docs_env/lib/python3.7/site-packages/keras/optimizer_v2/optimizer_
v2.py:356: UserWarning: The `lr` argument is deprecated, use `learning_rate` instead.
    "The `lr` argument is deprecated, use `learning_rate` instead.")
2023-09-12 11:23:04.815901: I tensorflow/stream_executor/cuda/cuda_dnn.cc:369]
Loaded cuDNN version 8100
2023-09-12 11:23:05.396630: I tensorflow/core/platform/default/subprocess.cc:304]
Start cannot spawn child process: No such file or directory
103/103 [==============================] - 7s 38ms/step - loss: 0.8676 - accuracy: 0.7618
Epoch 2/5
103/103 [==============================] - 4s 41ms/step - loss: 0.6568 - accuracy: 0.8880
Epoch 3/5
103/103 [==============================] - 4s 37ms/step - loss: 0.6238 - accuracy: 0.9111
Epoch 4/5
103/103 [==============================] - 4s 37ms/step - loss: 0.6009 - accuracy: 0.9245
Epoch 5/5
103/103 [==============================] - 4s 37ms/step - loss: 0.5872 - accuracy: 0.9287
```

(5) 评估模型，代码如下：

```
loss, accuracy = model.evaluate(test_data)
```

执行后会输出：

```
12/12 [==============================] - 2s 45ms/step - loss: 0.5993 - accuracy: 0.9292
```

（6）使用类 DataLoader 加载数据，代码如下：

```
data = DataLoader.from_folder(image_path)
```

假设同一个类的图像数据在同一个子目录下，子文件夹名就是类名。目前 DataLoader 支持的图像类型有 JPEG 格式和 PNG 格式。

然后将数据拆分为训练数据（80%）、验证数据（10%，可选）和测试数据（10%），代码如下：

```
train_data, rest_data = data.split(0.8)
validation_data, test_data = rest_data.split(0.5)
```

（7）输出显示 25 个带标签的图像例子，代码如下：

```
plt.figure(figsize=(10,10))
for i, (image, label) in enumerate(data.gen_dataset().unbatch().take(25)):
  plt.subplot(5,5,i+1)
  plt.xticks([])
  plt.yticks([])
  plt.grid(False)
  plt.imshow(image.numpy(), cmap=plt.cm.gray)
  plt.xlabel(data.index_to_label[label.numpy()])
plt.show()
```

执行效果如图 8-2 所示。

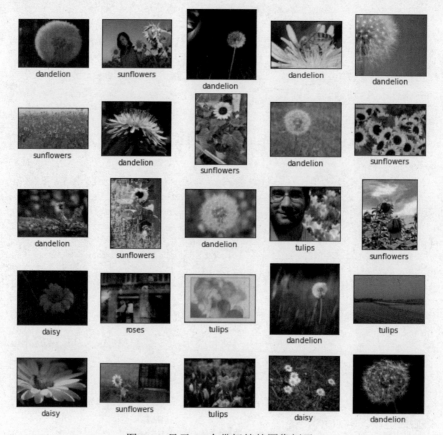

图 8-2　显示 25 个带标签的图像例子

(8)根据加载的数据创建自定义图像分类器模型,默认模型是 EfficientNet-Lite0。代码如下:

```
model = image_classifier.create(train_data, validation_data=validation_data)
```

(9)在 100 张测试图像中绘制预测结果,红色(方框标出)的预测标签是错误的预测结果,而其他是正确的。代码如下:

```
#如果预测结果与"测试"数据集中提供的标签不同,将以红色突出显示它。
plt.figure(figsize=(20, 20))
predicts = model.predict_top_k(test_data)
for i, (image, label) in enumerate(test_data.gen_dataset().unbatch().take(100)):
  ax = plt.subplot(10, 10, i+1)
  plt.xticks([])
  plt.yticks([])
  plt.grid(False)
  plt.imshow(image.numpy(), cmap=plt.cm.gray)

  predict_label = predicts[i][0][0]
  color = get_label_color(predict_label,
                          test_data.index_to_label[label.numpy()])
  ax.xaxis.label.set_color(color)
  plt.xlabel('Predicted: %s' % predict_label)
plt.show()
```

执行效果如图 8-3 所示。

图 8-3 在 100 张测试图像中绘制预测结果

8.2.2　将 Keras 模型转换为 TensorFlow Lite

经过前面的介绍，已经成功训练了数字分类器模型。在接下来的内容中，将这个模型转换为 TensorFlow Lite 格式并进行移动部署。

导出带有元数据的 TensorFlow Lite 模型，该元数据提供了模型描述的标准。标签文件嵌入在元数据中。默认的训练后量化技术是图像分类任务的全整数量化。导出代码如下：

```
model.export(export_dir='.')
```

执行后会输出：

```
2023-09-12 11:23:32.415723: I tensorflow/core/grappler/devices.cc:66] Number of eligible GPUs (core count >= 8, compute capability >= 0.0): 1
2023-09-12 11:23:32.415840: I tensorflow/core/grappler/clusters/single_machine.cc:357] Starting new session
2023-09-12 11:23:32.416303: I tensorflow/stream_executor/cuda/cuda_gpu_executor.cc:937] successful NUMA node read from SysFS had negative value (-1), but there must be at least one NUMA node, so returning NUMA node zero
2023-09-12 11:23:32.416699: I tensorflow/stream_executor/cuda/cuda_gpu_executor.cc:937] successful NUMA node read from SysFS had negative value (-1), but there must be at least one NUMA node, so returning NUMA node zero
2023-09-12 11:23:32.417007: I tensorflow/stream_executor/cuda/cuda_gpu_executor.cc:937] successful NUMA node read from SysFS had negative value (-1), but there must be at least one NUMA node, so returning NUMA node zero
2023-09-12 11:23:32.417414: I tensorflow/stream_executor/cuda/cuda_gpu_executor.cc:937] successful NUMA node read from SysFS had negative value (-1), but there must be at least one NUMA node, so returning NUMA node zero
2023-09-12 11:23:32.417738: I tensorflow/stream_executor/cuda/cuda_gpu_executor.cc:937] successful NUMA node read from SysFS had negative value (-1), but there must be at least one NUMA node, so returning NUMA node zero
2023-09-12 11:23:32.418047: I tensorflow/core/common_runtime/gpu/gpu_device.cc:1510] Created device /job:localhost/replica:0/task:0/device:GPU:0 with 14648 MB memory: -> device: 0, name: Tesla V100-SXM2-16GB, pci bus id: 0000:00:05.0, compute capability: 7.0
2023-09-12 11:23:32.451651: I tensorflow/core/grappler/optimizers/meta_optimizer.cc:1137] Optimization results for grappler item: graph_to_optimize
    function_optimizer: Graph size after: 913 nodes (656), 923 edges (664), time = 17.945ms.
    function_optimizer: function_optimizer did nothing. time = 0.391ms.

2023-09-12 11:23:33.380451: W tensorflow/compiler/mlir/lite/python/tf_tfl_flatbuffer_helpers.cc:351] Ignored output_format.
2023-09-12 11:23:33.380503: W tensorflow/compiler/mlir/lite/python/tf_tfl_flatbuffer_helpers.cc:354] Ignored drop_control_dependency.
2023-09-12    11:23:33.426653:    I tensorflow/compiler/mlir/tensorflow/utils/dump_mlir_util.cc:210] disabling MLIR crash reproducer, set env var `MLIR_CRASH_REPRODUCER_DIRECTORY` to enable.
fully_quantize: 0, inference_type: 6, input_inference_type: 3, output_inference_type: 3
WARNING:absl:For model inputs containing unsupported operations which cannot be quantized, the `inference_input_type` attribute will default to the original type.
INFO:tensorflow:Label file is inside the TFLite model with metadata.
INFO:tensorflow:Label file is inside the TFLite model with metadata.
INFO:tensorflow:Saving labels in /tmp/tmpny214hzn/labels.txt
INFO:tensorflow:Saving labels in /tmp/tmpny214hzn/labels.txt
INFO:tensorflow:TensorFlow Lite model exported successfully: ./model.tflite
INFO:tensorflow:TensorFlow Lite model exported successfully: ./model.tflite
```

上述模式可以集成到 Android 或使用 iOS 应用 ImageClassifier API 的 TensorFlow 精简版任

务库，允许的导出格式可以是以下之一：
- ExportFormat.TFLITE
- ExportFormat.LABEL
- ExportFormat.SAVED_MODEL

在默认情况下，只导出带有元数据的 TensorFlow Lite 模型。还可以有选择地导出不同的文件。例如下面的代码仅导出标签文件：

```
model.export(export_dir='.', export_format=ExportFormat.LABEL)
```

还可以使用该 evaluate_tflite 方法评估 tflite 模型，代码如下：

```
model.evaluate_tflite('model.tflite', test_data)
```

执行后会输出：

```
{'accuracy': 0.9019073569482289}
```

8.2.3 量化处理

接下来开始在 TensorFlow Lite 模型上自定义训练后量化。训练后量化是一种转换技术，可以简化模型和减少推理延迟，同时还可以提高 CPU 和硬件加速器的推理速度，但是模型的精度会略有下降。因此，它被广泛用于优化模型。

Model Maker 库在导出模型时应用默认的训练后量化技术，如果想自定义训练后量化，Model Maker 也支持使用 QuantizationConfig 的多个训练后量化选项。例如，下面的代码中我们以 float16 量化为例定义量化配置。

```
config = QuantizationConfig.for_float16()
```

然后用这样的配置导出 TensorFlow Lite 模型，代码如下：

```
model.export(export_dir='.', tflite_filename='model_fp16.tflite', quantization_config=config)

INFO:tensorflow:Assets written to: /tmp/tmp3tagi8ov/assets
INFO:tensorflow:Assets written to: /tmp/tmp3tagi8ov/assets
2023-09-12 11:33:19.486299: I tensorflow/stream_executor/cuda/cuda_gpu_executor.cc:937] successful NUMA node read from SysFS had negative value (-1), but there must be at least one NUMA node, so returning NUMA node zero
2023-09-12 11:33:19.486660: I tensorflow/core/grappler/devices.cc:66] Number of eligible GPUs (core count >= 8, compute capability >= 0.0): 1
2023-09-12 11:33:19.486769: I tensorflow/core/grappler/clusters/single_machine.cc:357] Starting new session
2023-09-12 11:33:19.487314: I tensorflow/stream_executor/cuda/cuda_gpu_executor.cc:937] successful NUMA node read from SysFS had negative value (-1), but there must be at least one NUMA node, so returning NUMA node zero
2023-09-12 11:33:19.487754: I tensorflow/stream_executor/cuda/cuda_gpu_executor.cc:937] successful NUMA node read from SysFS had negative value (-1), but there must be at least one NUMA node, so returning NUMA node zero
2023-09-12 11:33:19.488070: I tensorflow/stream_executor/cuda/cuda_gpu_executor.cc:937] successful NUMA node read from SysFS had negative value (-1), but there must be at least one NUMA node, so returning NUMA node zero
2023-09-12 11:33:19.488480: I tensorflow/stream_executor/cuda/cuda_gpu_executor.cc:937] successful NUMA node read from SysFS had negative value (-1), but there must be at least one NUMA node, so returning NUMA node zero
2023-09-12 11:33:19.488804: I tensorflow/stream_executor/cuda/cuda_gpu_executor.cc:937] successful NUMA node read from SysFS had negative value (-1), but there must be at least one NUMA node, so returning NUMA node zero
```

```
    2023-09-12  11:33:19.489094:   I   tensorflow/core/common_runtime/gpu/gpu_device.
cc:1510] Created device /job:localhost/replica:0/task:0/device:GPU:0 with 14648 MB memory:
-> device: 0, name: Tesla V100-SXM2-16GB, pci bus id: 0000:00:05.0, compute capability:
7.0
    2023-09-12 11:33:19.525503: I tensorflow/core/grappler/optimizers/meta_optimizer.
cc:1137] Optimization results for grappler item: graph_to_optimize
     function_optimizer: Graph size after: 913 nodes (656), 923 edges (664), time =
19.663ms.
     function_optimizer: function_optimizer did nothing. time = 0.423ms.
    INFO:tensorflow:Label file is inside the TFLite model with metadata.
    2023-09-12 11:33:19.358426: W tensorflow/compiler/mlir/lite/python/tf_tfl_flatbuffer_
helpers.cc:351] Ignored output_format.
    2023-09-12 11:33:19.358474: W tensorflow/compiler/mlir/lite/python/tf_tfl_flatbuffer_
helpers.cc:354] Ignored drop_control_dependency.
    INFO:tensorflow:Label file is inside the TFLite model with metadata.
    INFO:tensorflow:Saving labels in /tmp/tmpyiyio9gh/labels.txt
    INFO:tensorflow:Saving labels in /tmp/tmpyiyio9gh/labels.txt
    INFO:tensorflow:TensorFlow Lite model exported successfully: ./model_fp16.tflite
    INFO:tensorflow:TensorFlow Lite model exported successfully: ./model_fp16.tflite
```

8.2.4 更改模型

在创建模型后我们可以修改模型，具体来说可以通过如下几种方式进行修改。

1. 更改此库中支持的模型

我们创建的模型支持转换为 EfficientNet-Lite、MobileNetV2 和 ResNet50 模型，其中 EfficientNet-Lite 是一系列图像分类模型，可以实现先进的精度并适用于边缘设备。默认模型是 EfficientNet-Lite0。

只需在 create() 方法中将参数设置为 MobileNetV2 模型，就可以将模型切换到 MobileNetV2。代码如下：

```
model = image_classifier.create(train_data, model_spec=model_spec.get('mobilenet_v2'),
validation_data=validation_data)
```

执行后会输出：

```
    INFO:tensorflow:Retraining the models...
    INFO:tensorflow:Retraining the models...
    Model: "sequential_2"
    _____
    Layer (type)                 Output Shape              Param #
    =================================================================
    hub_keras_layer_v1v2_2 (HubK (None, 1280)              2257984

    dropout_2 (Dropout)          (None, 1280)              0

    dense_2 (Dense)              (None, 5)                 6405
    =================================================================
    Total params: 2,264,389
    Trainable params: 6,405
    Non-trainable params: 2,257,984
    _____
    None
    Epoch 1/5
    /tmpfs/src/tf_docs_env/lib/python3.7/site-packages/keras/optimizer_v2/optimizer_
v2.py:356: UserWarning: The `lr` argument is deprecated, use `learning_rate` instead.
      "The `lr` argument is deprecated, use `learning_rate` instead.")
```

```
    91/91 [==============================] - 8s 57ms/step - loss: 0.9474 - accuracy:
0.7486 - val_loss: 0.6713 - val_accuracy: 0.8807
    Epoch 2/5
    91/91 [==============================] - 5s 54ms/step - loss: 0.7013 - accuracy:
0.8764 - val_loss: 0.6342 - val_accuracy: 0.9119
    Epoch 3/5
    91/91 [==============================] - 5s 54ms/step - loss: 0.6577 - accuracy:
0.8963 - val_loss: 0.6328 - val_accuracy: 0.9119
    Epoch 4/5
    91/91 [==============================] - 5s 54ms/step - loss: 0.6245 - accuracy:
0.9176 - val_loss: 0.6445 - val_accuracy: 0.9006
    Epoch 5/5
    91/91 [==============================] - 5s 55ms/step - loss: 0.6034 - accuracy:
0.9303 - val_loss: 0.6290 - val_accuracy: 0.9091
```

评估新训练的 MobileNetV2 模型以查看测试数据的准确性和损失，代码如下：

```
loss, accuracy = model.evaluate(test_data)
```

执行后会输出：

```
    12/12 [==============================] - 1s 38ms/step - loss: 0.6723 - accuracy:
0.8883
```

2．更改 TensorFlow Hub 中的模型

还可以切换到其他新模型，输入图像并输出 TensorFlow Hub 格式的特征向量。以 Inception V3 模型为例，我们可以定义 inception_v3_specwhich 是 image_classifier.ModelSpec 的对象，包含 Inception V3 模型的规范。

我们需要指定模型名称 name、TensorFlow Hub 模型的 url uri。同时，默认值 input_image_shape 是[224, 224]，我们需要将其更改为[299, 299]。

```
inception_v3_spec = image_classifier.ModelSpec(
   uri='https://tfhub.dev/google/imagenet/inception_v3/feature_vector/1')
inception_v3_spec.input_image_shape = [299, 299]
```

然后，将参数 model_spec 设置为 inception_v3_specincreate，可以重新训练 Inception V3 模型。其余步骤完全相同，最终可以得到一个定制的 InceptionV3 TensorFlow Lite 模型。

3．更改自己的自定义模型

如果想使用 TensorFlow Hub 中没有的自定义模型，应该在 TensorFlow Hub 中创建和导出 ModelSpec。然后开始训练 ModelSpec，像上面的过程那样定义对象即可，不再赘述。

8.3 识别器的具体实现

在使用 TensorFlow 定义和训练机器学习模型，并将训练好的 TensorFlow 模型转换为 TensorFlow Lite 模型后，接下来将使用这个模型开发一个 Android 鲜花识别器系统。

8.3.1 准备工作

（1）使用 Android Studio 导入本项目源码"TFLClassify-main"，如图 8-4 所示。

（2）将 TensorFlow Lite 模型添加到工程。

将在之前训练的 TensorFlow Lite 模型文件 mnist.tflite 复制到 Android 工程下面的目录中：

```
TFLClassify-main/finish/src/main/ml
```

图 8-4 导入工程

（3）更新 build.gradle。

打开 app 模块中的文件 build.gradle，分别设置 Android 的编译版本和运行版本，设置需要使用的库文件，例如摄像头库 CameraX、GPU 代理库。最后添加对 TensorFlow Lite 模型库的引用。代码如下：

```
plugins {
    id 'com.android.application'
    id 'kotlin-android'

    //建议使用 Kotlin-kapt 进行数据绑定
    id 'kotlin-kapt'
}

android {
    compileSdkVersion 30

    defaultConfig {
        applicationId "org.tensorflow.lite.examples.classification"
        minSdkVersion 21
        targetSdkVersion 30
        versionCode 1
        versionName "1.0"

        testInstrumentationRunner "androidx.test.runner.AndroidJUnitRunner"
    }

    buildTypes {
        release {
            minifyEnabled false
            proguardFiles getDefaultProguardFile('proguard-android-optimize.txt'), 'proguard-rules.pro'
        }
    }

    //CameraX 需要 Java 8, 这个 compileOptions 块是必需的
    compileOptions {
        sourceCompatibility JavaVersion.VERSION_1_8
        targetCompatibility JavaVersion.VERSION_1_8
    }
    kotlinOptions {
        jvmTarget = '1.8'
```

```
    }
    //启用数据绑定
    buildFeatures{
        dataBinding = true
        mlModelBinding true
    }

}

dependencies {

    //Kotlin 和 Jetpack 的默认导入
    implementation "org.jetbrains.kotlin:kotlin-stdlib:$kotlin_version"
    implementation 'androidx.core:core-ktx:1.3.2'
    implementation 'androidx.appcompat:appcompat:1.2.0'
    implementation 'com.google.android.material:material:1.2.1'
    implementation 'org.tensorflow:tensorflow-lite-support:0.1.0-rc1'
    implementation "androidx.recyclerview:recyclerview:1.1.0"
    implementation 'org.tensorflow:tensorflow-lite-metadata:0.1.0-rc1'

    //导入 CameraX
    def camerax_version = "1.0.0-beta10"
    //使用 camera2 实现的 CameraX 核心库
    implementation "androidx.camera:camera-camera2:$camerax_version"
    //CameraX 生命周期库
    implementation "androidx.camera:camera-lifecycle:$camerax_version"
    //CameraX 视图类
    implementation "androidx.camera:camera-view:1.0.0-alpha17"
    implementation "androidx.activity:activity-ktx:1.1.0"

    // TODO 5: 可选 GPU 代理
    implementation 'org.tensorflow:tensorflow-lite-gpu:2.3.0'

    testImplementation 'junit:junit:4.13'
    androidTestImplementation 'androidx.test.ext:junit:1.1.2'
    androidTestImplementation 'androidx.test.espresso:espresso-core:3.3.0'
}
```

8.3.2 页面布局

本项目的页面布局文件是 activity_main.xml，功能是在 Android 界面中显示摄像机预览框视图，并在下方显示识别结果。文件 activity_main.xml 的具体实现代码如下：

```
<merge
    xmlns:android="http://schemas.android.com/apk/res/android"
    xmlns:app="http://schemas.android.com/apk/res-auto"
    xmlns:tools="http://schemas.android.com/tools"
    android:layout_width="match_parent"
    android:layout_height="match_parent"
    tools:context=".MainActivity">

    <androidx.camera.view.PreviewView
        android:id="@+id/viewFinder"
        android:layout_width="match_parent"
        android:layout_height="match_parent" />
```

```xml
            <androidx.appcompat.widget.Toolbar
                android:id="@+id/toolbar"
                android:layout_width="match_parent"
                android:layout_height="?attr/actionBarSize"
                android:layout_gravity="top"
                android:background="#8000">

                <ImageView
                    android:layout_width="wrap_content"
                    android:layout_height="wrap_content"
                    android:src="@drawable/tfl2_logo"
                    android:contentDescription="@string/tensorflow_lite_logo_description" />

            </androidx.appcompat.widget.Toolbar>
            <androidx.recyclerview.widget.RecyclerView
                android:id="@+id/recognitionResults"
                android:layout_width="match_parent"
                android:layout_height="wrap_content"
                android:layout_gravity="bottom"
                android:orientation="vertical"
                app:layoutManager="LinearLayoutManager"  />
    </merge>
```

上述代码在 RecyclerView 识别结果视图区域中调用 LinearLayoutManager 来显示识别结果。LinearLayoutManager 的功能在文件 recognition_item.xml 中通过两列文字显示识别结果。文件 recognition_item.xml 的具体实现代码如下：

```xml
<layout xmlns:android="http://schemas.android.com/apk/res/android"
    xmlns:tools="http://schemas.android.com/tools">

    <data>

        <variable
            name="recognitionItem"
            type="org.tensorflow.lite.examples.classification.viewmodel.Recognition" />
    </data>

    <LinearLayout
        android:layout_width="match_parent"
        android:layout_height="wrap_content"
        android:background="#8000"
        android:orientation="horizontal">

        <TextView
            android:id="@+id/recognitionName"
            android:layout_width="0dp"
            android:layout_height="wrap_content"
            android:layout_weight="2"
            android:padding="8dp"
            android:text="@{recognitionItem.label}"
            android:textColor="@color/white"
            android:textAppearance="?attr/textAppearanceHeadline6"
            tools:text="Orange" />

        <TextView
```

```xml
            android:id="@+id/recognitionProb"
            android:layout_width="0dp"
            android:layout_height="wrap_content"
            android:layout_weight="1"
            android:gravity="end"
            android:padding="8dp"
            android:text="@{recognitionItem.probabilityString}"
            android:textColor="@color/white"
            android:textAppearance="?attr/textAppearanceHeadline6"
            tools:text="99%" />

    </LinearLayout>
</layout>
```

8.3.3 实现 UI Activity

本项目的 UI Activity 功能是由文件 RecognitionAdapter.kt 实现，功能是使用项目布局和数据绑定来扩展 ViewHolder。文件 RecognitionAdapter.kt 的主要实现代码如下：

```kotlin
class RecognitionAdapter(private val ctx: Context) :
    ListAdapter<Recognition, RecognitionViewHolder>(RecognitionDiffUtil()) {

    /**
     * 使用项目布局和数据绑定来扩展 ViewHolder
     */
    override fun onCreateViewHolder(parent: ViewGroup, viewType: Int): RecognitionViewHolder {
        val inflater = LayoutInflater.from(ctx)
        val binding = RecognitionItemBinding.inflate(inflater, parent, false)
        return RecognitionViewHolder(binding)
    }

    //将数据字段绑定到 RecognitionViewHolder
    override fun onBindViewHolder(holder: RecognitionViewHolder, position: Int) {
        holder.bindTo(getItem(position))
    }

    private class RecognitionDiffUtil : DiffUtil.ItemCallback<Recognition>() {
        override fun areItemsTheSame(oldItem: Recognition, newItem: Recognition): Boolean {
            return oldItem.label == newItem.label
        }

        override fun areContentsTheSame(oldItem: Recognition, newItem: Recognition): Boolean {
            return oldItem.confidence == newItem.confidence
        }
    }

}

class RecognitionViewHolder(private val binding: RecognitionItemBinding) :
    RecyclerView.ViewHolder(binding.root) {
```

```
        //将所有字段绑定到视图（要查看哪个UI元素绑定到哪个字段），请查看文件layout/recognition_
item.xml
        fun bindTo(recognition: Recognition) {
            binding.recognitionItem = recognition
            binding.executePendingBindings()
        }
    }
```

8.3.4 实现主Activity

本项目的主 Activity 功能是由文件 MainActivity.kt 实现，功能是调用前面的布局文件 activity_main.xml 在屏幕上方显示一个摄像机预览界面，并在屏幕下方显示识别结果的文字信息。文件 MainActivity.kt 的具体实现流程如下：

（1）定义需要的常量，设置在屏幕中显示 3 行预测，设置使用摄像机权限。代码如下：

```
//常量
private const val MAX_RESULT_DISPLAY = 3      //显示的最大结果数
private const val TAG = "TFL Classify"         //日志记录的名称
private const val REQUEST_CODE_PERMISSIONS = 999 //获取请求权限
private val REQUIRED_PERMISSIONS = arrayOf(Manifest.permission.CAMERA)//摄像机权限

//ImageAnalyzer结果的侦听器
typealias RecognitionListener = (recognition: List<Recognition>) -> Unit
```

（2）创建 TensorFlow Lite 分类器的入口类 MainActivity，打开摄像机预览功能，并在下方实现实时识别。代码如下：

```
class MainActivity : AppCompatActivity() {

    // CameraX 变量
    private lateinit var preview: Preview //预览实例，快速、灵敏地查看摄像机
    private lateinit var imageAnalyzer: ImageAnalysis // 分析实例，用于运行ML代码
    private lateinit var camera: Camera
    private val cameraExecutor = Executors.newSingleThreadExecutor()

    //视图附件
    private val resultRecyclerView by lazy {
        findViewById<RecyclerView>(R.id.recognitionResults) //显示分析结果
    }
    private val viewFinder by lazy {
        findViewById<PreviewView>(R.id.viewFinder) //显示来自摄像机的预览图像
    }

    //识别结果。因为它是一个viewModel，所以它可以在屏幕旋转后继续使用
    private val recogViewModel: RecognitionListViewModel by viewModels()

    override fun onCreate(savedInstanceState: Bundle?) {
        super.onCreate(savedInstanceState)
        setContentView(R.layout.activity_main)

        //请求摄像机权限
        if (allPermissionsGranted()) {
            startCamera()
        } else {
            ActivityCompat.requestPermissions(
                this, REQUIRED_PERMISSIONS, REQUEST_CODE_PERMISSIONS
```

```
        )
    }

    //初始化 resultRecyclerView 及其连接的 ViewAdapter
    val viewAdapter = RecognitionAdapter(this)
    resultRecyclerView.adapter = viewAdapter

    // 禁用"回放视图"动画以减少闪烁,否则项目会随着列表的更改而移动、淡入和淡出
    resultRecyclerView.itemAnimator = null

    // 在 recognitionList 的 LiveData 字段上附加一个观察者
    //每当在 recognitionList 的 LiveData 字段上设置新列表时,将通知 recycler 视图进行更新
    recogViewModel.recognitionList.observe(this,
        Observer {
            viewAdapter.submitList(it)
        }
    )
}
```

(3)编写函数 allPermissionsGranted(),在本例中的功能是检查是否获取操作摄像机的权限。代码如下:

```
private fun allPermissionsGranted(): Boolean = REQUIRED_PERMISSIONS.all {
    ContextCompat.checkSelfPermission(
        baseContext, it
    ) == PackageManager.PERMISSION_GRANTED
}
```

(4)编写函数 onRequestPermissionsResult(),功能是弹出是否开启"摄像机权限"提醒框窗口。代码如下:

```
override fun onRequestPermissionsResult(
    requestCode: Int,
    permissions: Array<String>,
    grantResults: IntArray
) {
    if (requestCode == REQUEST_CODE_PERMISSIONS) {
        if (allPermissionsGranted()) {
            startCamera()
        } else {
            // 如果未授予权限,请退出应用程序
            //更多有关权限信息的说明,请参阅:
            // https://developer.android.com/training/permissions/usage-notes
            Toast.makeText(
                this,
                getString(R.string.permission_deny_text),
                Toast.LENGTH_SHORT
            ).show()
            finish()
        }
    }
}
```

(5)编写函数 startCamera(),功能是启动手机中的摄像机,具体包括如下 4 个功能:
- 初始化预览用例;
- 初始化图像分析仪用例;
- 将上述两者都附加到此活动的生命周期;

- 通过管道将预览对象的输出传输到屏幕上的 PreviewView 视图。

函数 startCamera() 的具体实现代码如下：

```kotlin
private fun startCamera() {
    val cameraProviderFuture = ProcessCameraProvider.getInstance(this)

    cameraProviderFuture.addListener(Runnable {
        //将摄像机的生命周期绑定到生命周期所有者
        val cameraProvider: ProcessCameraProvider = cameraProviderFuture.get()

        preview = Preview.Builder()
            .build()

        imageAnalyzer = ImageAnalysis.Builder()
            //为要分析的图像设置了理想的尺寸，CameraX 将选择可能不完全相同或保持相同纵横比的最合适的分辨率
            .setTargetResolution(Size(224, 224))
            // 图像分析仪应如何输入：1.每帧，但不掉帧；2.转到最新帧，可能会丢失一些帧。默认值为2
            // STRATEGY_KEEP_ONLY_LATEST。以下行是可选的，为了清晰起见保留在此处
            .setBackpressureStrategy(ImageAnalysis.STRATEGY_KEEP_ONLY_LATEST)
            .build()
            .also { analysisUseCase: ImageAnalysis ->
                analysisUseCase.setAnalyzer(cameraExecutor, ImageAnalyzer(this) { items ->
                    //更新已识别对象的列表
                    recogViewModel.updateData(items)
                })
            }

        // 选择"摄像机"，默认为"后退"。如果不可用，请选择前摄像头
        val cameraSelector =
            if (cameraProvider.hasCamera(CameraSelector.DEFAULT_BACK_CAMERA))
                CameraSelector.DEFAULT_BACK_CAMERA else CameraSelector.DEFAULT_FRONT_CAMERA

        try {
            // 在重新绑定之前解除绑定实例
            cameraProvider.unbindAll()

            //将实例绑定到摄像机（尝试一次绑定所有内容），CameraX 将找到最佳组合
            camera = cameraProvider.bindToLifecycle(
                this, cameraSelector, preview, imageAnalyzer
            )

            // 将预览附加到预览视图，也称为取景器
            preview.setSurfaceProvider(viewFinder.surfaceProvider)
        } catch (exc: Exception) {
            Log.e(TAG, "Use case binding failed", exc)
        }

    }, ContextCompat.getMainExecutor(this))
}
```

（6）编写类 ImageAnalyzer，功能是分析摄像机中采集的图片信息，使用 TensorFlow Lite 模型实现图像识别。代码如下：

```kotlin
    private class ImageAnalyzer(ctx: Context, private val listener: Recognition-
Listener) :
        ImageAnalysis.Analyzer {

        // TODO 1：添加类变量TensorFlow Lite模型
        //通过lazy初始化flowerModel，以便在调用process方法时它在同一线程中运行
        private val flowerModel: FlowerModel by lazy{

            // TODO 6.可选选项，开启GPU加速
            val compatList = CompatibilityList()

            val options = if(compatList.isDelegateSupportedOnThisDevice) {
                Log.d(TAG, "This device is GPU Compatible ")
                Model.Options.Builder().setDevice(Model.Device.GPU).build()
            } else {
                Log.d(TAG, "This device is GPU Incompatible ")
                Model.Options.Builder().setNumThreads(4).build()
            }

            //初始化花模型
            FlowerModel.newInstance(ctx, options)
        }

        override fun analyze(imageProxy: ImageProxy) {

            val items = mutableListOf<Recognition>()

            // TODO 2：将图像转换为位图，然后转换为TensorImage
            val tfImage = TensorImage.fromBitmap(toBitmap(imageProxy))

            // TODO 3：使用经过训练的模型对图像进行处理，并对处理结果进行排序和挑选
            val outputs = flowerModel.process(tfImage)
                .probabilityAsCategoryList.apply {
                    sortByDescending { it.score } //首先以最高的得分排序
                }.take(MAX_RESULT_DISPLAY) //以最高的得分为例

            // TODO 4：将最高概率项转换为识别列表
            for (output in outputs) {
                items.add(Recognition(output.label, output.score))
            }

            //返回结果
            listener(items.toList())

            // 关闭图像，告诉CameraX将下一个图像提供给分析仪
            imageProxy.close()
        }

        /**
         * 将图像转换为位图
         */
        private val yuvToRgbConverter = YuvToRgbConverter(ctx)
        private lateinit var bitmapBuffer: Bitmap
        private lateinit var rotationMatrix: Matrix

        @SuppressLint("UnsafeExperimentalUsageError")
```

```kotlin
            private fun toBitmap(imageProxy: ImageProxy): Bitmap? {

                val image = imageProxy.image ?: return null

                //初始化缓冲区
                if (!::bitmapBuffer.isInitialized) {
                    //图像旋转和RGB图像缓冲区仅初始化一次
                    Log.d(TAG, "Initalise toBitmap()")
                    rotationMatrix = Matrix()
                    rotationMatrix.postRotate(imageProxy.imageInfo.rotationDegrees.toFloat())
                    bitmapBuffer = Bitmap.createBitmap(
                        imageProxy.width, imageProxy.height, Bitmap.Config.ARGB_8888
                    )
                }

                //将图像传递给图像分析器
                yuvToRgbConverter.yuvToRgb(image, bitmapBuffer)

                //以正确的方向创建位图
                return Bitmap.createBitmap(
                    bitmapBuffer,
                    0,
                    0,
                    bitmapBuffer.width,
                    bitmapBuffer.height,
                    rotationMatrix,
                    false
                )
            }
        }
    }
```

8.3.5 图像转换

编写文件YuvToRgbConverter.kt，功能是将YUV_420_888格式的数据转换为RGB对象。YUV即通过Y、U和V三个分量表示颜色空间，其中Y表示亮度，U和V表示色度。不同于RGB中每个像素点都有独立的R、G和B三个颜色分量值，YUV根据U和V采样数目的不同，分为如YUV444、YUV422和YUV420等，而YUV420表示的就是每个像素点有一个独立的亮度表示，即Y分量；而色度则由每4个像素点共享一个。举例来说，对于4×4的图片，在YUV420下，有16个Y值、4个U值和4个V值。

YUV420根据颜色数据的存储顺序不同，又分为多种不同的格式，如YUV420Planar、YUV420PackedPlanar、YUV420SemiPlanar和YUV420PackedSemiPlanar，这些格式实际存储的信息是完全一致的。举例来说，对于4×4的图片，在YUV420下，任何格式都有16个Y值、4个U值和4个V值，不同格式只是Y、U和V的排列顺序变化。I420（YUV420Planar的一种）则为：

YYYYYYYYYYYYYYYYUUUUVVVV

NV21（YUV420SemiPlanar）则为：

YYYYYYYYYYYYYYYYUVUVUVUV

也就是说，YUV420是一类格式的集合，YUV420并不能完全确定颜色数据的存储顺序。

对于 YUV 来说图片的宽和高是必不可少的，因为 YUV 本身只存储颜色信息，想要还原出图片，必须知道图片的宽和高。在 Android 中，使用 Image 保存有图片的宽和高，这可以分别通过函数 getWidth()和 getHeight()得到。每个 Image 有自己的格式，这个格式由 ImageFormat 确定。对于 YUV420 来说，ImageFormat 在 API>=21 的 Android 系统中新加入了 YUV_420_888 类型，其表示 YUV420 格式的集合，888 表示 Y、U、V 分量中每个颜色占 8bit。既然只能指定 YUV420 这个格式集合，那怎么知道具体的格式呢？Y、U 和 V 三个分量的数据分别保存在三个 Plane 类中，可以通过 getPlanes()得到。Plane 实际是对 ByteBuffer 的封装。Image 保证了 plane #0 一定是 Y，#1 一定是 U，#2 一定是 V。且对于 plane #0，Y 分量数据一定是连续存储的，中间不会有 U 或 V 数据穿插，也就是说我们一定能够一次性得到所有 Y 分量的值。

文件 YuvToRgbConverter.kt 的具体实现流程如下：

```kotlin
class YuvToRgbConverter(context: Context) {
    private val rs = RenderScript.create(context)
    private val scriptYuvToRgb = ScriptIntrinsicYuvToRGB.create(rs, Element.U8_4(rs))

    private var pixelCount: Int = -1
    private lateinit var yuvBuffer: ByteBuffer
    private lateinit var inputAllocation: Allocation
    private lateinit var outputAllocation: Allocation

    @Synchronized
    fun yuvToRgb(image: Image, output: Bitmap) {

        //确保在已分配的输出缓冲区范围内进行处理
        if (!::yuvBuffer.isInitialized) {
            pixelCount = image.cropRect.width() * image.cropRect.height()
            //每个像素位是整个图像的平均值，因此计算完整缓冲区的大小是非常有用的，但不应用于确定像素偏移
            val pixelSizeBits = ImageFormat.getBitsPerPixel(ImageFormat.YUV_420_888)
            yuvBuffer = ByteBuffer.allocateDirect(pixelCount * pixelSizeBits / 8)
        }

        //回退缓冲区，不需要清除它，因为它将被填充
        yuvBuffer.rewind()

        //使用 NV21 格式获取字节数组形式的 YUV 数据
        imageToByteBuffer(image, yuvBuffer.array())

        //确保已分配 RenderScript 输入和输出
        if (!::inputAllocation.isInitialized) {
            //显式创建一个 NV21 类型的元素，因为这是我们使用的像素格式
            val elemType = Type.Builder(rs, Element.YUV(rs)).setYuvFormat(ImageFormat.NV21).create()
            inputAllocation = Allocation.createSized(rs, elemType.element, yuvBuffer.array().size)
        }
        if (!::outputAllocation.isInitialized) {
            outputAllocation = Allocation.createFromBitmap(rs, output)
        }

        //将 NV21 格式 YUV 转换为 RGB
        inputAllocation.copyFrom(yuvBuffer.array())
        scriptYuvToRgb.setInput(inputAllocation)
        scriptYuvToRgb.forEach(outputAllocation)
        outputAllocation.copyTo(output)
    }
```

```kotlin
private fun imageToByteBuffer(image: Image, outputBuffer: ByteArray) {
    if (BuildConfig.DEBUG && image.format != ImageFormat.YUV_420_888) {
        error("Assertion failed")
    }

    val imageCrop = image.cropRect
    val imagePlanes = image.planes

    imagePlanes.forEachIndexed { planeIndex, plane ->
        // 输入时需要为每个输出值设置读取多少个值,仅 Y 平面为每个像素设置一个值,U 和 V 的分辨率为一半,即:
        // Y Plane            U Plane     V Plane
        // ===============    =======     =======
        // Y Y Y Y Y Y Y Y    U U U U     V V V V
        // Y Y Y Y Y Y Y Y    U U U U     V V V V
        // Y Y Y Y Y Y Y Y    U U U U     V V V V
        // Y Y Y Y Y Y Y Y    U U U U     V V V V
        // Y Y Y Y Y Y Y Y
        // Y Y Y Y Y Y Y Y
        // Y Y Y Y Y Y Y Y
        val outputStride: Int

        //写入输出缓冲区中的索引的下一个值,对于 Y 来说它是零,对于 U 和 V 来说从 Y 的末尾开始并交叉处理
        //
        // First chunk        Second chunk
        // ===============    ===============
        // Y Y Y Y Y Y Y Y    V U V U V U V U
        // Y Y Y Y Y Y Y Y    V U V U V U V U
        // Y Y Y Y Y Y Y Y    V U V U V U V U
        // Y Y Y Y Y Y Y Y    V U V U V U V U
        // Y Y Y Y Y Y Y Y
        // Y Y Y Y Y Y Y Y
        // Y Y Y Y Y Y Y Y
        var outputOffset: Int

        when (planeIndex) {
            0 -> {
                outputStride = 1
                outputOffset = 0
            }
            1 -> {
                outputStride = 2
                //对于 NV21 格式,U 为奇数索引
                outputOffset = pixelCount + 1
            }
            2 -> {
                outputStride = 2
                //对于 NV21 格式,V 是偶数索引
                outputOffset = pixelCount
            }
            else -> {
                //图像包含 3 个以上的平面
                return@forEachIndexed
            }
        }

        val planeBuffer = plane.buffer
        val rowStride = plane.rowStride
        val pixelStride = plane.pixelStride
```

```kotlin
            //如果不是Y平面,必须将宽度和高度除以2
            val planeCrop = if (planeIndex == 0) {
                imageCrop
            } else {
                Rect(
                    imageCrop.left / 2,
                    imageCrop.top / 2,
                    imageCrop.right / 2,
                    imageCrop.bottom / 2
                )
            }

            val planeWidth = planeCrop.width()
            val planeHeight = planeCrop.height()

            //用于存储每行字节的中间缓冲区
            val rowBuffer = ByteArray(plane.rowStride)

            //每行的大小(字节)
            val rowLength = if (pixelStride == 1 && outputStride == 1) {
                planeWidth
            } else {
                // 因为步幅可以包括来自除该特定平面和行之外的像素的数据,并且该数据可以在像素之间,而不是在每个像素之后:
                //
                // |---- Pixel stride ----|                    Row ends here --> |
                // | Pixel 1 | Other Data | Pixel 2 | Other Data | ... | Pixel N |
                //
                //|----像素跨距--|行结束于此-->|
                //
                //|像素1 |其他数据|像素2 |其他数据|……|像素N|
                ////我们需要得到(N-1)*(像素步幅字节)每行+1字节的最后一个像素
                (planeWidth - 1) * pixelStride + 1
            }

            for (row in 0 until planeHeight) {
                //将缓冲区位置移到此行的开头
                planeBuffer.position(
                    (row + planeCrop.top) * rowStride + planeCrop.left * pixelStride
                )

                if (pixelStride == 1 && outputStride == 1) {
                    //当像素和输出有一个步长值时,我们可以在一个步长中复制整行
                    planeBuffer.get(outputBuffer, outputOffset, rowLength)
                    outputOffset += rowLength
                } else {
                    //当像素或输出的跨距大于1时,我们必须逐像素复制
                    planeBuffer.get(rowBuffer, 0, rowLength)
                    for (col in 0 until planeWidth) {
                        outputBuffer[outputOffset] = rowBuffer[col * pixelStride]
                        outputOffset += outputStride
                    }
                }
            }
        }
    }
}
```

到此为止,整个项目工程全部开发完毕。单击 Android Studio 顶部的运行按钮运行本项目,在 Android 设备中将会显示执行效果,如图 8-5 所示。在屏幕上方会显示摄像头的拍摄界面,在下方显示摄像头视频的识别结果。

图 8-5 执行效果

8.3.6 使用 GPU 委托加速

TensorFlow Lite 支持多种硬件加速器,以加快移动设备上的推理速度。其中 GPU 是 TensorFlow Lite 可以通过委托机制利用的加速器之一,它非常易于使用。在本项目中模块下的 build.gradlestart 文件中,添加了如下的依赖:

```
implementation 'org.tensorflow:tensorflow-lite-gpu:2.3.0'
```

我们可以在通过 Android Studio 导入 TensorFlow Lite 时,在"Import"界面中选择"Auto add TensorFlow Lite"复选框的方式实现同样的功能,这样即可启用 GPU 加速功能,如图 8-6 所示。

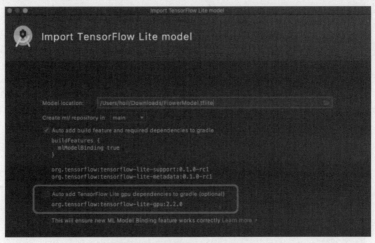

图 8-6 启用 GPU 加速功能

第 9 章　智能素描绘图系统开发

智能绘画系统在人工智能领域比较容易常见，很多人认为人工智能技术的出现会替代绘画者和设计师。本章开发了一个基于图像处理和绘图技术的系统，旨在实现自动化图像绘制。该系统使用了一系列算法和技术，包括边缘增强、图像量化和调色以及笔画绘制。该系统可以将输入图像转换为以笔画形式绘制的艺术作品。

9.1　背景介绍

本项目开发的是一款经典的智能素描绘图系统软件，旨在提供一个有趣、创造性的平台，让用户能够轻松地绘制各种素描图像。无论是专业画家还是初学者，都可以通过本项目绘制出精美的素描图片。该系统具有以下主要特点：

- 绘画工具：软件提供了各种绘画工具，包括不同颜色的画笔、油漆、纹理和装饰品。用户可以根据自己的喜好选择合适的工具，绘制出个性化的素描图像。
- 多种绘画模式：提供了多种绘画模式，如自由绘画模式、填色模式和涂鸦模式。用户可以根据自己的创意选择不同的模式，创作出独一无二的艺术品。
- 保存和分享：用户可以将绘制的图像保存到本地设备，并分享给朋友、社交媒体或打印成实体作品。
- 用户友好界面：软件具有简洁直观的用户界面，易于操作和导航。无论是专业艺术家还是非专业人士，都能够轻松上手并享受绘画的乐趣。

9.2　需求分析

1. 功能需求
- 图形绘制：允许用户选择绘制的图形类型（如人物、景色、猫、狗、树等）并指定相应的参数（如大小、颜色、位置等），以绘制该图形并显示在屏幕上。
- 图形编辑：允许用户选择已绘制的图形，并对其进行编辑操作，如修改颜色、调整大小、移动位置等。
- 图形删除：允许用户选择已绘制的图形，并将其从画布上删除。
- 图形保存：允许用户将绘制的图形保存为图像文件，以便后续使用或分享。
- 图形加载：允许用户从图像文件中加载已保存的图形，并显示在画布上进行编辑或查看。

2. 用户界面需求
- 图形选择界面：提供一个用户界面，展示可供选择的图形类型列表，用户可以从中选择要绘制的图形。
- 参数输入界面：在图形选择后，显示相应的参数输入界面，允许用户输入图形的参数值，如大小、颜色、位置等。

- 画布显示界面：提供一个画布区域，用于显示绘制的图形，以及进行编辑和删除操作。
- 文件操作界面：提供文件操作选项，允许用户保存绘制的图形为图像文件或加载已保存的图形文件。

3．性能需求

- 响应时间：软件应在用户输入命令或进行操作后能够快速响应，并及时显示绘制的图形或执行相应的操作。
- 稳定性：软件应具备稳定性和健壮性，能够处理各种异常情况和错误输入，并提供相应的提示和处理方式。
- 可扩展性：软件应具备良好的可扩展性，方便后续增加更多图形类型和功能。

4．安全性需求

- 用户身份验证：如有需要，软件应提供用户身份验证机制，确保只有授权用户可以进行绘制和编辑操作。
- 数据保护：软件应确保用户输入的参数和绘制的图形数据得到适当的保护，防止数据泄露或篡改。

5．兼容性需求

- 平台兼容性：软件应兼容多个操作系统平台，如 Windows、macOS 和 Linux 等。
- 文件格式兼容性：软件应支持常见的图像文件格式，如 PNG、JPEG 等，以方便图形的保存和加载。

9.3 功能模块

本项目的具体结构如图 9-1 所示。

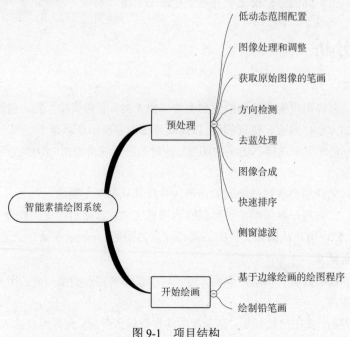

图 9-1　项目结构

9.4 预处理

在正式绘图之前，需要预先对图像进行处理，主要包括低动态范围配置、图像处理和调整、获取原始图像的笔画、方向检测、去蓝处理、图像合成、快速排序和侧窗滤波。本节详细讲解这些预处理功能的实现过程。

9.4.1 低动态范围配置

编写文件 LDR.py 实现图像的低动态范围配置，LDR 常见的格式有 png 和 jpg 等，比如一张 8 位图，RGB 三个通道灰度值变化范围 0~255，即 8 位图能表示 256^3 种颜色，看似很多，但实际上远不够来描述我们身处真实世界中的色彩。文件 LDR.py 的具体实现流程如下：

（1）编写方法 LDR() 实现图像的低动态范围处理，功能是将输入的图像转换为浮点数格式，将像素值划分为 n 个区间，然后进行量化，并将像素值重新映射回原始范围。对应的实现代码如下：

```python
def LDR(img, n):
    Interval = 2510.0/n
    img = np.float32(img)
    img = np.uint8(img/Interval)
    img = np.clip(img,0,n-1)
    img = np.uint8((img+0.5)*Interval)
    return img
```

（2）编写方法 HistogramEqualization()，功能是调用函数 createCLAHA() 生成自适应均衡化图像，参数 clipLimit 表示颜色对比度的阈值，参数 titleGridSize 用于设置像素均衡化的网格大小，即在多少网格下进行直方图的均衡化操作。方法 HistogramEqualization() 的具体实现代码如下：

```python
def HistogramEqualization(img,clipLimit=2, tileGridSize=(10,10)):
    # 创建一个CLAHE对象（参数是可选的）
    clahe = cv2.createCLAHE(clipLimit=clipLimit, tileGridSize=tileGridSize)
    img = clahe.apply(img)
    return img
```

（3）然后在测试主函数中加载一个灰度图像(img_path = './input/jiangwen/010s.jpg')，对图像进行直方图均衡化处理(img = HistogramEqualization(img))。具体实现代码如下：

```python
if __name__ == '__main__':
    img_path = './input/jiangwen/010s.jpg'
    img = cv2.imread(img_path, cv2.IMREAD_GRAYSCALE)

    img = HistogramEqualization(img)

    s = SideWindowFilter(radius=1, iteration=1)
    img_tensor = F.to_tensor(img).unsqueeze(0)

    res = s.forward(img_tensor).squeeze().detach().numpy()

    for n in range(8, 11):
        img_ldr = LDR(res, n)
        cv2.imwrite("LDR{}.jpg".format(n), img_ldr)

    print("done")
```

对上述代码的具体说明如下：

- 首先，创建一个 SideWindowFilter 对象(s = SideWindowFilter(radius=1, iteration=1))。
- 将图像转换为 PyTorch 张量(img_tensor = F.to_tensor(img).unsqueeze(0))。
- 使用 SideWindowFilter 对象对图像进行滤波处理(res = s.forward(img_tensor).squeeze().detach().numpy())。
- 对处理后的图像进行多次 LDR 处理，并将结果保存为不同的图像文件(for n in range(8, 11): img_ldr = LDR(res, n) cv2.imwrite("LDR{}.jpg".format(n), img_ldr))。
- 打印"done"表示处理完成。

9.4.2 图像处理和调整

编写文件 tone.py 对图像进行处理和调整，主要实现图像色调调整和单通道低动态范围处理的功能。文件 tone.py 的具体实现流程如下：

（1）导入所需的库，包括 cv2、numpy 和 math。具体实现代码如下：

```
import cv2
import numpy as np
import math
```

（2）定义函数 transferTone()，用于调整图像的色调。首先，计算输入图像的直方图，并对直方图进行归一化。然后，根据给定的权重和概率密度函数，计算每个像素的新值，以实现色调调整。最后，将调整后的图像显示并返回。函数 transferTone()的具体实现代码如下：

```
def transferTone(img):
    ho, _ = np.histogram(img.flatten(), bins=256, range=[0, 256], density=True)
    ho_cumulative = np.cumsum(ho)

    omiga1 = 76
    omiga2 = 22
    omiga3 = 2
    p1 = lambda x: (1 / 9.0) * np.exp(-(255 - x) / 9.0)
    p2 = lambda x: (1.0 / (225 - 105)) * ((x >= 105) & (x <= 225))
    p3 = lambda x: (1.0 / np.sqrt(2 * math.pi * 11)) * np.exp(-((x - 90) ** 2) / float((2 * (11 ** 2))))
    p = lambda x: (omiga1 * p1(x) + omiga2 * p2(x) + omiga3 * p3(x)) * 0.01

    prob = np.array([p(i) for i in range(256)])
    prob /= np.sum(prob)
    histo_cumulative = np.cumsum(prob)

    Iadjusted = np.interp(ho_cumulative, histo_cumulative, np.arange(256))
    Iadjusted = np.uint8(Iadjusted.reshape(img.shape))

    cv2.imshow('adjust tone', Iadjusted)
    cv2.waitKey(0)
    J = cv2.blur(Iadjusted, (3, 3))
    cv2.imshow('blurred adjust tone', J)
    cv2.waitKey(1)
    return J
```

（3）创建函数 LDR_single()，用于对图像进行单通道的低动态范围（LDR）处理。首先，将输入图像分为 n 个区间，然后对每个区间的像素进行量化，并将像素值映射为对应的色调。最后，将每个区间的色调图像保存到指定路径下。函数 LDR_single()的具体实现代码如下：

```python
def LDR_single(img, n, output_pathos):
    Interval = 250.0 / n
    img = np.float32(img)
    img = np.uint8(img / Interval)
    img = np.clip(img, 0, n - 1)

    tones = [(i + 0.5) * Interval for i in range(n)]
    tones = np.array(tones)
    mask = np.zeros(img.shape, dtype=bool)
    for i in range(n):
        mask = (img == i)
        tone = np.uint8(tones[i] * mask + (1 - mask) * 255)
        cv2.imwrite(output_pathos + "/tone{}.png".format(i), tone)

    return
```

（4）创建函数 LDR_single_add()，用于对图像进行单通道的累积低动态范围（LDR）处理。与 LDR_single 函数类似，它将输入图像分为 n 个区间，并根据累积的像素数量计算每个像素的色调值。同时，还将每个区间的二值掩膜和累积的掩膜保存到指定路径下。函数 LDR_single_add() 的具体实现代码如下：

```python
def LDR_single_add(img, n, output_pathos):
    Interval = 250.0 / n
    img = np.float32(img)
    img = np.uint8(img / Interval)
    img = np.clip(img, 0, n - 1)

    mask_add = np.zeros(img.shape, dtype=bool)
    for i in range(n):
        mask = (img == i)
        mask_add += mask
        cv2.imwrite(output_pathos + "/mask/mask{}.png".format(i), np.uint8(mask_add * 255))
        tone = np.uint8((i + 0.5) * Interval * mask_add + (1 - mask_add) * 255)
        cv2.imwrite(output_pathos + "/mask/tone_cumulate{}.png".format(i), tone)

    return
```

（5）加载一个灰度图像(img_path = './input/jw.png')，调用 LDR_single 函数对图像进行单通道的 LDR 处理，并将结果保存到指定路径下(LDR_single(img, 10, output_pathos="./output"))。调用 LDR_single_add 函数对图像进行单通道的累积 LDR 处理，并将结果保存到指定路径下(LDR_single_add(img, 10, output_pathos="./output"))。具体实现代码如下：

```python
if __name__ == '__main__':
    img_path = './input/jw.png'
    img = cv2.imread(img_path, cv2.IMREAD_GRAYSCALE)

    # img = transferTone(img)
    # cv2.imwrite("./input/jiangwen/transferTone.png", img)

    LDR_single(img, 10, output_pathos="./output")
    LDR_single_add(img, 10, output_pathos="./output")
    print("done")
```

9.4.3 获取原始图像的笔画

在计算机视觉和图像处理领域，"stroke"通常指代图像中的笔画或轮廓。获取图像的笔画可以涉及边缘检测、轮廓提取或图像分割等技术。这些技术可以帮助我们找到图像中物体的边界或轮廓,并进一步分析和处理这些边界信息。要计算和获取图像的笔画,可以使用各种图像处理库和算法。常用的包括 OpenCV、Pillow、scikit-image 等,具体的实现方法会根据应用的需求和所使用的库而有所不同。

在本项目中,编写文件 Stroke_origin.py 计算和获取原始图像的笔画(stroke)。通过计算图像的梯度信息和滤波操作,生成具有线条效果的笔画图像,并对图像进行后处理以增强视觉效果。文件 Stroke_origin.py 的具体实现流程如下:

(1)编写函数 zhuanImg(),功能是对输入的图像进行旋转操作,返回旋转后的图像。具体实现代码如下:

```python
def zhuanImg(img, angle):
    # 将图像按给定角度旋转
    row, col = img.shape
    M = cv2.getRotationMatrix2D((row / 2, col / 2), angle, 1)
    res = cv2.warpAffine(img, M, (row, col))
    return res
```

(2)编写函数 genStroke(),功能是接收一个灰度图像和 dirNum 参数作为输入,生成图像的笔画效果。具体实现代码如下:

```python
# 计算和获取原始图像的笔画
def genStroke(img, dirNum, verbose=False):
    height, width = img.shape[0], img.shape[1]
    img = np.float32(img) / 2510.0
    print("输入图像高度: %d, 宽度: %d" % (height, width))

    print("图像预处理,去噪...")
    img = cv2.medianBlur(img, 3)

    print("生成梯度图像...")
    imX = np.append(np.absolute(img[:, 0:width - 1] - img[:, 1:width]), np.zeros((height, 1)), axis=1)
    imY = np.append(np.absolute(img[0:height - 1, :] - img[1:height, :]), np.zeros((1, width)), axis=0)
    ################################################################
    ##### 有许多方法来生成渐变效果 #####
    ################################################################
    img_gradient = np.sqrt((imX ** 2 + imY ** 2))
    img_gradient = imX + imY
    if verbose == True:
        cv2.imshow('梯度图像', np.uint8(255 - img_gradient * 255))
        cv2.imwrite('output/grad.jpg', np.uint8(255 - img_gradient * 255))
        cv2.waitKey(0)

    # 滤波核大小
    tempsize = 0
    if height > width:
        tempsize = width
    else:
        tempsize = height
    tempsize /= 30
    ################################################################
    #核大小是边长的 1/30
    ################################################################
    halfKsize = int(tempsize / 2)
    if halfKsize < 1:
```

```python
        halfKsize = 1
    if halfKsize > 9:
        halfKsize = 9
    kernalsize = halfKsize * 2 + 1
    print("核大小 = %s" % (kernalsize))

    ################################################################
    ############### 这里生成滤波核 ###################
    ################################################################
    kernel = np.zeros((dirNum, kernalsize, kernalsize))
    kernel[0, halfKsize, :] = 1.0
    for i in range(0, dirNum):
        kernel[i, :, :] = temp = zhuanImg(kernel[0, :, :], i * 180 / dirNum)
        kernel[i, :, :] *= kernalsize / np.sum(kernel[i])
        if verbose == True:
            title = '线性滤波核 %d' % i
            cv2.imshow(title, np.uint8(temp * 255))
            cv2.waitKey(0)

    ##########################################################
    # cv2.filter2D() 其实做的是 correlate 而不是 conv
    # correlate 相当于 kernel 旋转180° 的 conv
    # 但是我们的 kernel 是中心对称的,所以不影响
    ##########################################################

    # 在不同方向上滤波梯度图像
    print("在不同方向上滤波梯度图像...")
    response = np.zeros((dirNum, height, width))
    for i in range(dirNum):
        ker = kernel[i, :, :]
        response[i, :, :] = cv2.filter2D(img_gradient, -1, ker)
    if verbose == True:
        for i in range(dirNum):
            title = '响应 %d' % i
            cv2.imshow(title, np.uint8(response[i, :, :] * 255))
            cv2.waitKey(0)

    # 将梯度图像分成不同的子图
    print("计算梯度分类...")
    Cs = np.zeros((dirNum, height, width))
    for x in range(width):
        for y in range(height):
            i = np.argmax(response[:, y, x])
            Cs[i, y, x] = img_gradient[y, x]
    if verbose == True:
        for i in range(dirNum):
            title = '最大响应 %d' % i
            cv2.imshow(title, np.uint8(Cs[i, :, :] * 255))
            cv2.waitKey(0)

    # 生成线条形状
    print("生成线条形状...")
    spn = np.zeros((dirNum, height, width))
    for i in range(dirNum):
        ker = kernel[i, :, :]
        spn[i, :, :] = cv2.filter2D(Cs[i], -1, ker)
    sp = np.sum(spn, axis=0)

    sp = sp * np.power(img_gradient, 0.4)
```

```
        sp = (sp - np.min(sp)) / (np.max(sp) - np.min(sp))
        S = 1 - sp
        return S
```

上述代码的执行流程如下：

① 对输入图像进行预处理，包括去噪声操作（使用 cv2.medianBlur 函数）。

② 生成图像的梯度信息（imX 和 imY），计算梯度幅值（img_gradient）。

③ 生成线性核（kernel），并对其进行旋转和归一化处理。

④ 使用生成的核对图像的梯度进行滤波操作，得到不同方向上的响应值（response）。

⑤ 将梯度图像根据响应值进行分类，得到不同方向上的梯度子图（Cs）。

⑥ 通过对子图进行滤波操作得到形状线（spn），并将所有子图求和得到最终的笔画图像（sp）。

⑦ 对笔画图像进行后处理，调整其亮度和对比度，并进行归一化处理得到最终的笔画图像（S）。

（3）加载输入图像并调用函数 genStroke() 生成笔画图像，对生成的笔画图像进行处理，包括指数变换（stroke=np.power(stroke, 3)）和归一化处理。最后，将最终的笔画图像保存到文件并显示出来。具体实现代码如下：

```python
    if __name__ == '__main__':

        img_path = './input/1.jpg'
        img = cv2.imread(img_path, cv2.IMREAD_GRAYSCALE)
        stroke = genStroke(img, 18, False)
        # stroke = stroke*(np.exp(stroke)-np.exp(1)+1)
        stroke = np.power(stroke, 3)
        # stroke=(stroke - np.min(stroke)) / (np.max(stroke) - np.min(stroke))  # 加深边缘
        stroke = np.uint8(stroke * 255)

        cv2.imwrite('output/edge.jpg', stroke)
        cv2.imshow('笔画', stroke)
        cv2.waitKey(0)
```

代码执行后会得到图像 cat.jpg 的原始图画，如图 9-2 所示。

图 9-2　图像 cat.jpg 的原始图画

9.4.4　方向检测

图像的方向检测在计算机视觉和图像处理领域中具有多种应用。下面是一些常见的用途：

- 图像分类与识别：在图像分类和识别任务中，了解图像中的主要方向可以提供有关图像内容和结构的重要信息。主要方向可以用于特征提取、特征匹配和模式识别等任务，有助于改善分类和识别的准确性。
- 边缘检测与分割：图像的主要方向可以用于边缘检测和图像分割任务。通过检测图像中的边缘和线条，可以更好地理解图像中的结构和边界。方向信息可以用于优化边缘检测算法和分割方法，从而提高结果的质量和准确性。
- 图像增强与滤波：了解图像中的主要方向可以帮助改善图像的质量和清晰度。根据主要方向的信息，可以应用方向性滤波器或增强算法，以增强图像的特定方向上的特征和细节，从而改善图像的视觉效果。
- 视觉导航与目标跟踪：在计算机视觉中，方向信息对于视觉导航和目标跟踪非常重要。通过检测和跟踪图像中的主要方向，可以帮助机器视觉系统进行定位、导航和目标追踪，从而实现自主导航、目标跟踪和场景理解等任务。

在本项目中，通过编写文件 tools.py 对输入的图像进行方向检测，并输出图像中的主要方向。文件 tools.py 的具体实现流程如下：

（1）编写函数 get_start_end(mask)，根据给定的二值掩码图像获取图像中连续的区间的起始和结束位置。函数 get_start_end(mask)的具体实现代码如下：

```python
def get_start_end(mask):
    lines=[]
    Flag = True  # 没有新的间隔
    for i in range(mask.shape[0]):
        if Flag == True:
            if mask[i]==1:
                if len(lines)>0 and i-lines[-1][1]<=1:  ####### 太接近了
                    Flag = False
                    continue
                else:
                    lines.append([i,i])
                    Flag = False
            else:
                continue
        else:
            if mask[i]==1:
                continue
            else:
                lines[-1][1]=i
                Flag = True

    if Flag == False:
        lines[-1][1]=i

    return lines
```

（2）编写函数 zhuanImg(img, angle)，功能是对给定图像进行旋转，旋转角度由参数 angle 指定。函数 zhuanImg(img, angle)的具体实现代码如下：

```python
def zhuanImg(img, angle):
    row, col = img.shape
    M  = cv2.getRotationMatrix2D((row / 2 , col / 2 ), angle, 1)
    res = cv2.warpAffine(img, M, (row, col))
    return res
```

（3）编写函数 get_directions()，功能是在给定的图像中检测出主要的方向。其中，Num_choose 表示要选择的主要方向数量，dirNum 表示在方向空间中划分的方向数量，img 是输入的灰度图像。在函数 get_directions()中，首先对输入图像计算梯度图像，并根据梯度图像生成掩码。然后，构建滤波核，并将滤波核应用于梯度图像，得到在不同方向上滤波后的响应图像。接着，根据响应图像的最大值确定每个像素点的主要方向，并统计各个方向的像素数量。最后，根据像素数量选择主要的方向，并返回这些方向的角度值。函数 get_directions()的具体实现代码如下：

```python
def get_directions(Num_choose, dirNum, img):
    height,width = img.shape
    img = np.float32(img)/2510.0
    # print("输入图像高度: %d，宽度: %d" % (height,width))
    
    imX = np.append(np.absolute(img[:, 0 : width - 1]  - img[:, 1 : width]), np.zeros((height, 1)), axis = 1)
    imY = np.append(np.absolute(img[0 : height - 1, :] - img[1 : height, :]), np.zeros((1, width)), axis = 0)
    
    img_gradient = np.sqrt((imX ** 2 + imY ** 2))
    mask = (img_gradient-0.02)>0
    cv2.imshow('mask',np.uint8(mask*255))
    # img_gradient = imX + imY
    
    # 滤波核大小
    tempsize = 0
    if height > width:
        tempsize = width
    else:
        tempsize = height
    tempsize /= 30  # 根据论文，核大小是边长的1/30
    
    halfKsize = int(tempsize / 2)
    if halfKsize < 1:
        halfKsize = 1
    if halfKsize > 9:
        halfKsize = 9
    kernalsize = halfKsize * 2 + 1
    # print("核大小 = %s" %(kernalsize))
    
    ################################################################
    ############### 这里生成滤波核 ##################
    ################################################################
    kernel = np.zeros((dirNum, kernalsize, kernalsize))
    kernel [0,halfKsize,:] = 1.0
    for i in range(0,dirNum):
        kernel[i,:,:] = temp = zhuanImg(kernel[0,:,:], i * 180 / dirNum)
        kernel[i,:,:] *= kernalsize/np.sum(kernel[i])
    
    # 在不同方向上滤波梯度图像
    print("在不同方向上滤波梯度图像 ...")
    response = np.zeros((dirNum, height, width))
    for i in range(dirNum):
        ker = kernel[i,:,:];
        response[i, :, :] = cv2.filter2D(img_gradient, -1, ker)
```

```
    cv2.waitKey(0)

# 将梯度图像分成不同的子图
print("计算方向分类 ...")
direction = np.zeros(( height, width))
for x in range(width):
    for y in range(height):
        direction[y, x] = np.argmax(response[:,y,x])
# direction = direction*mask

dirs = np.zeros(dirNum)
for i in range (dirNum):
    dirs[i]=np.sum((direction-i)==0)
sort_dirs = np.sort(dirs,axis=0)
print(dirs,sort_dirs)
angles = []
for i in range(Num_choose):
    for j in range (dirNum):
        if sort_dirs[-1-i]==dirs[j]:
            angles.append(j*180/dirNum)
            continue
return angles
```

（4）读取指定的灰度图像 cat.jpg，然后调用函数 get_directions()检测图像中的主要方向，并将结果保存在 angles 变量中。具体实现代码如下：

```
if __name__ == '__main__':
    input_pathos = './input/cat.jpg'
    img = cv2.imread(input_pathos, cv2.IMREAD_GRAYSCALE)
    angles = get_directions(4,12,img)
```

执行后会对图像 cat.jpg 进行方向检测，并输出图像中的主要方向：

```
在不同方向上滤波梯度图像 ...
计算方向分类 ...
[13849. 27010. 3283. 2941. 3641. 1400. 6787. 1122. 3174. 3480.
 3416. 4827.] [ 1122. 1400. 27010. 2941. 3174. 3283. 3416. 3480. 3641. 4827.
 6787. 13849.]
```

9.4.5 去蓝处理

去蓝处理是一种图像处理技术，主要目的是减少或消除图像中的蓝色成分。这种处理可以用于不同的应用和场景，主要用途如下：

- 色彩校正：某些图像可能受到蓝色光线的干扰或偏色，通过去除图像中的蓝色成分，可以校正图像的色彩偏差，使其更加真实和准确。
- 特定效果：去蓝处理可以产生一种老化或怀旧的效果，通过减少蓝色成分，使图像呈现出较为暖色调的效果，增强了图像的古典或复古感。
- 图像增强：在某些情况下，蓝色成分可能对图像的细节和对比度产生负面影响。通过去除蓝色成分，可以减少图像中的噪声或干扰，提高图像的清晰度和细节。
- 图像分析：对于一些特定的图像分析任务，蓝色成分可能对目标或特征的识别造成干扰。通过去除蓝色成分，可以减少干扰，提高目标或特征的检测和分析准确性。

需要注意的是，在具体应用中是否需要去蓝处理取决于实际需求和图像的特点。有些情况下可能需要保留蓝色成分，而有些情况下则需要去除。去蓝处理作为一种图像处理技术，可根据具体需求进行调整和应用。

在本项目中编写文件 dellblue.py 实现去蓝处理功能，具体实现流程如下：

（1）定义函数 deblue()用于对图像进行去蓝处理。在函数内部，首先将输入的灰度图像转换为彩色图像（BGR 格式）。接着，将彩色图像转换为 HSV 颜色空间。使用高斯函数模拟生成两个高斯噪声图像，分别用于修改 HSV 图像的色调（H 通道）和饱和度（S 通道）。然后将修改后的 HSV 图像转换回 BGR 颜色空间，并将其转换为 8 位无符号整数类型，将处理后的图像保存到指定路径下的文件 aging.jpg 中。函数 deblue()的具体实现代码如下：

```python
def dellblue(img, output_pathos):
    BGR = cv2.cvtColor(img, cv2.COLOR_GRAY2BGR)
    HSV = cv2.cvtColor(BGR, cv2.COLOR_BGR2HSV)

    size = img.shape

    HSV[:,:,0] = Gassian(size, mean=15, var=1)
    HSV[:,:,1] = Gassian(size, mean=20, var=2)

    result = cv2.cvtColor(HSV, cv2.COLOR_HSV2BGR)
    result = np.uint8(result)

    cv2.imwrite(output_pathos+'/aging.jpg', result)
    cv2.imshow("aging",result)
    cv2.waitKey(0)
```

（2）读取输入图像 "./output/draw.png" 的路径，然后调用函数 deblue()对输入图像进行去蓝处理，并将处理后的图像保存到指定路径下的文件中。具体实现代码如下：

```python
if __name__ == '__main__':
    input_pathos = './output/draw.png'
    output_pathos = './output'
    input_img = cv2.imread(input_pathos, cv2.IMREAD_GRAYSCALE)
    dellblue(input_img, output_pathos)
```

9.4.6 图像合成

在前面的去蓝处理文件 dellblue.py 中调用了文件 simulate.py 的内容，文件 simulate.py 实现了一个简单的图像合成功能，通过生成随机的平行直线并在画布上叠加，产生一幅具有一定曲线和衰减效果的图像。文件 simulate.py 的具体实现流程如下：

（1）编写函数 Gassian(size, mean, var)，功能是生成一个服从高斯分布的随机矩阵，用于添加噪声。具体实现代码如下：

```python
constant_length = 1000

def Gassian(size, mean = 0, var = 0):
    norm = np.random.randn(*size)
    denorm = norm * np.sqrt(var) + mean
    return np.uint8(np.round(np.clip(denorm,0,255)))
```

在上述代码中，constant_length = 1000 是一个常数，用于判断直线的长度是否达到阈值。在代码中，当直线的长度小于 constant_length 时，直线是对齐的（Aligned），即直线的起点和终点在同一水平位置上。当直线的长度大于等于 constant_length 时，直线不再对齐（Not Aligned），即直线的起点和终点不在同一水平位置上。这个阈值的设置主要用于确定直线的对齐方式。如果长度较短，直线对齐可以使得整体效果更加规整；而长度较长时，不对齐的直线可以产生一种更加自由和随机的效果。

（2）编写函数 Getline(distribution, length)，功能是根据给定的长度和分布参数生成一条随机直线。根据直线长度 constant_length 的大小，选择对齐和不对齐的方式生成直线。这个阈值的设定可以根据具体需求进行调整，以获得期望的直线效果。具体实现代码如下：

```
def Getline(distribution, length):
    linewidth = distribution.shape[0]
    if length < constant_length: # if length is too short, lines are Aligned
        patch = Gassian((2*linewidth, length), mean=250, var=3)
        for i in range(linewidth):
            patch[i] = Gassian((1, length), mean=distribution[i, 0], var=distribution[i, 1])
        begin, end = 0, 1
    else: # if length is't too short, lines is't Aligned
        patch = Gassian((2*linewidth, length+4*linewidth), mean=250, var=3)
        begin = np.clip(np.round(2.0 * linewidth), 0, 4 * linewidth).astype(np.uint8)
        end = np.clip(np.round(2.0 * linewidth), 1, 4 * linewidth + 1).astype(np.uint8)
        real_length = length + 4 * linewidth - end - begin
        patch[:linewidth, begin:-end] = np.array([Gassian((1, real_length), mean=distribution[i, 0], var=distribution[i, 1]) for i in range(linewidth)])

    patch = Attenuation(patch, linewidth=linewidth, distribution=distribution, begin=begin, end=end)
    patch = Distortion(patch, begin=begin, end=end)

    return np.clip(patch, 0, 255).astype(np.uint8)
```

（3）编写函数 Attenuation(patch, linewidth, distribution, begin, end)，功能是对输入的图像进行衰减处理，使图像在水平方向上逐渐变淡。具体实现代码如下：

```
def Attenuation(patch, linewidth, distribution, begin, end):
    order = int((patch.shape[1]-begin-end)/2)+1
    radius = (linewidth-1)/2
    canvas = Gassian((patch.shape[0], patch.shape[1]), mean=250, var=3)
    patch = np.float32(patch)
    canvas = np.float32(canvas)
    for i in range(begin, patch.shape[1]-end+1):
        for j in range(linewidth):
            a = np.abs((1.0-(i-begin)/order)**2)/3
            b = np.abs((1.0-j/radius)**2)*1
            patch[j,i] += (canvas[j,i]-patch[j,i])*np.sqrt(a+b)/1.5
            # patch[j,i] += 0.75*(canvas[j,i]-patch[j,i]) * (np.abs((1.0-(i-begin)/order)**2))**0.5

    return np.uint8(np.round(np.clip(patch,0,255)))
```

（4）编写函数 Distortion(patch, begin, end)，功能是对输入的图像进行扭曲处理，使图像呈现一定的曲线形状。具体实现代码如下：

```
def Distortion(patch, begin, end):
    height = patch.shape[0] // 2
    length = patch.shape[1]
    patch = patch.astype(np.float32)
    patch_copy = patch.copy()

    central = (length - begin - end) / 2 + begin
    if length > 100:
        radius = length ** 2 / (4 * height)
```

```
        else:
            radius = length ** 2 / (2 * height)

        offset_vals = ((central - np.arange(length)) ** 2) / (2 * radius)
        int_offsets = offset_vals.astype(int)
        decimal_offsets = offset_vals - int_offsets

        for i in range(length):
            int_offset = int_offsets[i]
            decimal_offset = decimal_offsets[i]
            for j in range(height):
                if j > int_offset:
                    patch[j, i] = int(decimal_offset * patch_copy[j - 1 - int_offset, i]
+ (1 - decimal_offset) * patch_copy[j - int_offset, i])
                else:
                    patch[j, i] = np.random.randn() * np.sqrt(3) + 250

        patch_copy = patch.copy()
        if length > 100:
            for i in range(length):
                int_offset = int_offsets[i]
                decimal_offset = decimal_offsets[i]
                for j in range(patch.shape[0]):
                    if j > int_offset:
                        patch[j, i] = int(decimal_offset * patch_copy[j - 1 - int_offset,
i] + (1 - decimal_offset) * patch_copy[j - int_offset, i])
                    else:
                        patch[j, i] = np.random.randn() * np.sqrt(3) + 250

        return np.clip(patch, 0, 255).astype(np.uint8)
```

（5）编写函数 GetParallel(distribution, height, length, linewidth)，功能是根据给定的分布参数、高度、长度和线宽生成一组平行的随机直线。具体实现代码如下：

```
def GetParallel(distribution, height, length, linewidth):
    if length<constant_length: # constant length
        canvas = Gassian((height+2*linewidth,length), mean=250, var = 3)
    else: # variable length
        canvas = Gassian((height+2*linewidth,length+4*linewidth), mean=250, var = 3)

    distensce = Gassian((1,int(height/linewidth)+2), mean = linewidth, var =
linewidth/5)
    # distensce = Gassian((1,int(height/linewidth)+1), mean = linewidth, var = 0)
    distensce = np.uint8(np.round(np.clip(distensce, linewidth*0.8,linewidth*1.25)))

    begin = 0
    for i in np.squeeze(distensce).tolist():
        newline = Getline(distribution=distribution, length=length)
        h,w = newline.shape
        # cv2.imshow('line', newline)
        # cv2.waitKey(0)
        # cv2.imwrite("D:/ECCV2020/simu_patch/Line3.jpg",newline)

        if begin < height:
            m = np.minimum(canvas[begin:(begin + h),:], newline)
            canvas[begin:(begin + h),:] = m
            begin += i
        else:
```

```
            break

    return canvas[:height,:]
```

(6)编写函数 ChooseDistribution(linewidth, Grayscale),功能是根据线宽和灰度值选择合适的分布参数,用于生成直线的灰度分布。具体实现代码如下:

```
def ChooseDistribution(linewidth, Grayscale):
    distribution = np.zeros((linewidth,2))
    c = linewidth/2.0
    difference = 250-Grayscale
    for i in range(distribution.shape[0]):
        distribution[i][0] = Grayscale + difference*abs(i-c)/c
        distribution[i][1] = np.cos((i-c)/c*(0.5*3.1415929))*difference

    return np.abs(distribution)
```

(7)在主程序中,首先生成一个基础画布 canvas,然后根据给定的参数调用函数来生成随机的平行直线,并将生成的直线放置在画布上。最后将显示生成的图像,并将其保存为 "maomao.png"。具体实现代码如下:

```
if __name__ == '__main__':
    np.random.seed(100)
    canvas = Gassian((500,500), mean=250, var = 3)

    # distribution = np.array([[245,31],[238,27],[218,48],[205,33],[214,38],[234,24],
[240,42]])
    linewidth = 7
    Grayscale = 128
    H,L = (100,200)
    distribution = ChooseDistribution(linewidth=linewidth, Grayscale=Grayscale)
    print(distribution)
    patch = GetParallel(distribution=distribution, height=H, length=L, linewidth=
linewidth)
    (h,w) = patch.shape
    # patch = GetOffsetParallel(offset=4, distribution=distribution, patch_size=
(40,200), linewidth_mean=distribution.shape[0], linewidth_var=1)
    # (h,w) = patch.shape
    canvas[250-int(h/2):250-int(h/2)+h,250-int(w/2):250-int(w/2)+w] = patch

    # cv2.imshow('Parallel', patch[:, 2*distribution.shape[0]:w-2*distribution.
shape[0]])
    cv2.imshow('Parallel', canvas)
    cv2.waitKey(0)
    cv2.imwrite("maomao.png",patch)
    print("done")
```

9.4.7 快速排序

编写文件 quicksort.py 实现快速排序算法,用于对给定数组进行排序。具体实现流程如下:

(1)编写函数 partition(),功能是根据选取的随机基准元素将数组分割为两部分,并返回基准元素的索引位置。函数内部使用了双指针的方式进行元素交换,将小于基准的元素放在基准的左边,大于基准的元素放在基准的右边。具体实现代码如下:

```
def partition(arr, low, high):
    # 随机选择基准元素
    random_index = random.randint(low, high)
```

```
            arr[random_index], arr[high] = arr[high], arr[random_index]

            pivot = arr[high]
            i = low - 1

            for j in range(low, high):
                if arr[j]['importance'] >= pivot['importance']:
                    i = i + 1
                    arr[i], arr[j] = arr[j], arr[i]

            arr[i+1], arr[high] = arr[high], arr[i+1]
            return i + 1
```

（2）编写函数 quickSort()，功能是通过递归对子数组进行快速排序。函数首先检查子数组的长度是否小于等于阈值，若是，则使用插入排序进行排序。若子数组长度大于阈值，则选择一个基准元素，将子数组划分为两部分，并通过递归对这两部分进行排序。为了优化性能，在递归调用中，会对较短的子数组进行尾递归优化，以减少递归深度。具体实现代码如下：

```
        def quickSort(arr, low, high):
            # 设置阈值，当子数组长度小于等于该阈值时，使用插入排序
            threshold = 10

            while low < high:
                # 当子数组长度小于等于阈值时，使用插入排序
                if high - low + 1 <= threshold:
                    insertionSort(arr, low, high)
                    return

                pi = partition(arr, low, high)

                # 对较短的子数组进行尾递归优化
                if pi - low < high - pi:
                    quickSort(arr, low, pi - 1)
                    low = pi + 1
                else:
                    quickSort(arr, pi + 1, high)
                    high = pi - 1
```

（3）编写函数 insertionSort()，功能是实现插入排序算法，用于在子数组长度小于等于阈值时进行排序。该函数通过比较元素的重要性（importance）进行排序，将重要性较高的元素向前移动，以实现递减排序。具体实现代码如下：

```
        def insertionSort(arr, low, high):
            for i in range(low + 1, high + 1):
                key = arr[i]
                j = i - 1

                while j >= low and arr[j]['importance'] < key['importance']:
                    arr[j + 1] = arr[j]
                    j -= 1

                arr[j + 1] = key
```

9.4.8　侧窗滤波

编写文件 SideWindowFilter.py 定义一个名为 SideWindowFilter 的 PyTorch 模型，用于实现

侧窗滤波。侧窗滤波是一种图像处理技术，通过在每个像素周围的邻域内进行滤波操作来改变图像的外观。文件 SideWindowFilter.py 的具体实现代码如下：

```python
import torch
import torch.nn as nn
import torch.nn.functional as F

class SideWindowFilter(nn.Module):

    def __init__(self, radius, iteration, filter='box'):
        super(SideWindowFilter, self).__init__()
        self.radius = radius
        self.iteration = iteration
        self.kernel_size = 2 * self.radius + 1
        self.filter = filter

    def forward(self, im):
        b, c, h, w = im.size()

        d = torch.zeros(b, 8, h, w, dtype=torch.float)
        res = im.clone()

        if self.filter.lower() == 'box':
            filter = torch.ones(1, 1, self.kernel_size, self.kernel_size)
            L, R, U, D = [filter.clone() for _ in range(4)]

            L[:, :, :, self.radius + 1:] = 0
            R[:, :, :, 0: self.radius] = 0
            U[:, :, self.radius + 1:, :] = 0
            D[:, :, 0: self.radius, :] = 0

            NW, NE, SW, SE = U.clone(), U.clone(), D.clone(), D.clone()

            L, R, U, D = L / ((self.radius + 1) * self.kernel_size), R / ((self.radius + 1) * self.kernel_size), \
                         U / ((self.radius + 1) * self.kernel_size), D / ((self.radius + 1) * self.kernel_size)

            NW[:, :, :, self.radius + 1:] = 0
            NE[:, :, :, 0: self.radius] = 0
            SW[:, :, :, self.radius + 1:] = 0
            SE[:, :, :, 0: self.radius] = 0

            NW, NE, SW, SE = NW / ((self.radius + 1) ** 2), NE / ((self.radius + 1) ** 2), \
                             SW / ((self.radius + 1) ** 2), SE / ((self.radius + 1) ** 2)

            # sum = self.kernel_size * self.kernel_size
            # sum_L, sum_R, sum_U, sum_D, sum_NW, sum_NE, sum_SW, sum_SE = \
            #         (self.radius + 1) * self.kernel_size, (self.radius + 1) * self.kernel_size, \
            #         (self.radius + 1) * self.kernel_size, (self.radius + 1) * self.kernel_size, \
            #         (self.radius + 1) ** 2, (self.radius + 1) ** 2, (self.radius + 1) ** 2, (self.radius + 1) ** 2

            print('L:', L)
            print('R:', R)
            print('U:', U)
            print('D:', D)
            print('NW:', NW)
            print('NE:', NE)
```

```
                print('SW:', SW)
                print('SE:', SE)

            for ch in range(c):
                im_ch = im[:, ch, ::].clone().view(b, 1, h, w)
                # print('im size in each channel:', im_ch.size())

                for i in range(self.iteration):
                    # print('###', (F.conv2d(input=im_ch, weight=L, padding=(self.radius,
self.radius)) / sum_L -
                    # im_ch).size(), d[:, 0,::].size())
                    d[:, 0, ::] = F.conv2d(input=im_ch, weight=L, padding=(self.radius,
self.radius)) - im_ch
                    d[:, 1, ::] = F.conv2d(input=im_ch, weight=R, padding=(self.radius,
self.radius)) - im_ch
                    d[:, 2, ::] = F.conv2d(input=im_ch, weight=U, padding=(self.radius,
self.radius)) - im_ch
                    d[:, 3, ::] = F.conv2d(input=im_ch, weight=D, padding=(self.radius,
self.radius)) - im_ch
                    d[:, 4, ::] = F.conv2d(input=im_ch, weight=NW, padding=(self.radius,
self.radius)) - im_ch
                    d[:, 5, ::] = F.conv2d(input=im_ch, weight=NE, padding=(self.radius,
self.radius)) - im_ch
                    d[:, 6, ::] = F.conv2d(input=im_ch, weight=SW, padding=(self.radius,
self.radius)) - im_ch
                    d[:, 7, ::] = F.conv2d(input=im_ch, weight=SE, padding=(self.radius,
self.radius)) - im_ch

                    d_abs = torch.abs(d)
                    print('im_ch', im_ch)
                    print('dm = ', d_abs.shape, d_abs)
                    mask_min = torch.argmin(d_abs, dim=1, keepdim=True)
                    print('mask min = ', mask_min.shape, mask_min)
                    dm = torch.gather(input=d, dim=1, index=mask_min)
                    im_ch = dm + im_ch

                res[:, ch, ::] = im_ch
            return res
```

该模型的构造函数__init__()接受参数 radius、iteration 和 filter，分别用于设置滤波器的半径、迭代次数和滤波类型。函数 forward()定义了模型的前向传播过程，接受输入图像 im，并根据设置的滤波类型进行侧窗滤波操作。

在上述代码中还使用了库 torch 的一些功能，如 torch.zeros、torch.ones、torch.clone、torch.view、torch.abs、torch.argmin 和 torch.gather 等。此外，还使用了 torch.nn.functional 中的 F.conv2d 函数进行卷积操作。另外，还使用了 PIL 和 OpenCV 库来加载和显示图像，以及使用 NumPy 进行数据处理。

这段代码实现了侧窗滤波的功能，通过定义模型和实现前向传播过程，对输入图像进行侧窗滤波操作，并输出滤波后的图像。

9.5 开始绘图

经过前面的预处理工作之后，接下来开始步入正式绘图工作。本节详细讲解两种绘图方案的实现过程。

9.5.1 基于边缘绘画的绘图程序

编写文件 cat.py，功能是实现一个基于边缘绘画的图像艺术生成算法，将输入的图像转换为具有艺术效果的图像。文件 cat.py 的实现步骤如下：

（1）基于输入路径和输出路径设置参数，包括一些控制绘画风格和过程的变量。
（2）对输入图像进行预处理，包括图像路径处理和创建输出路径。
（3）运行边缘细化滤波器（ETF）算法，生成边缘图像。
（4）对输入图像进行色调转换和直方图均衡化等处理。
（5）计算图像梯度，并进行归一化处理。
（6）将图像进行量化，生成低动态范围（LDR）图像。
（7）根据绘画方向和灰度级别，选择合适的笔触参数，形成笔触序列。
（8）根据笔触序列，在绘画画布上逐步绘制图像。
（9）定期保存绘画过程的中间结果图像。
（10）根据绘画结果生成边缘图像。
（11）将绘画结果与边缘图像进行合并，生成最终的结果图像。
（12）对结果图像进行颜色处理，去除蓝色成分。
（13）将结果图像与原始彩色图像合成，生成最终的彩色图像。
（14）将生成的图像保存到输出路径。

文件 cat.py 实现了一种基于边缘绘画的图像艺术生成算法，通过组合边缘细化、量化、笔触绘制和颜色处理等步骤，将输入图像转换为艺术效果的图像。具体实现流程如下：

（1）设置一系列参数，用于控制整个图像处理和绘制的过程。通过对这些参数的设置，可以影响最终生成的手绘风格图像的效果和绘制过程的展示方式。通过调整这些参数，可以控制线条的样式、绘制顺序以及图像处理的方式，从而获得不同的艺术化效果。具体实现代码如下：

```
input_pathos = './input/cat_up.png'
output_pathos = './output'

np.random.seed(1)
n = 6
linewidth = 4
direction = 10
Freq = 100
deepen = 1
transTone = False
kernel_radius = 3
iter_time = 15
background_dir = 45
CLAHE = True
edge_CLAHE = True
draw_new = True
random_order = False
ETF_order = True
process_visible = True
```

每个参数的功能说明如下：

- input_pathos：输入图像的路径。

- output_pathos：输出结果的路径。
- np.random.seed(1)：设置随机数种子，用于保持随机结果的可重复性。
- n：量化顺序，用于将灰度值分为多个级别。
- linewidth：线条的宽度。
- direction：绘制线条的方向数量。
- Freq：保存绘制结果的频率，表示每绘制多少条线条保存一次结果。
- deepen：用于边缘加深的参数。
- transTone：是否进行色调转换。
- kernel_radius：边缘切向流（ETF）算法中的核半径。
- iter_time：ETF 算法的迭代次数。
- background_dir：ETF 算法中的背景方向。
- CLAHE：是否进行直方图自适应均衡化。
- edge_CLAHE：是否对边缘图像进行直方图自适应均衡化。
- draw_new：是否重新绘制线条。
- random_order：是否对线条绘制顺序进行随机排序。
- ETF_order：是否按照 ETF 算法结果的顺序绘制线条。
- process_visible：是否显示绘制过程的中间结果。

（2）根据输入路径中的文件名创建相应的输出文件夹，并在输出文件夹中创建两个子文件夹"mask"和"process"，用于存储相关的结果和中间过程。这样可以更好地组织和保存处理后的图像及其相关数据。具体实现代码如下：

```
file_name = os.path.basename(input_pathos)
file_name = file_name.split('.')[0]
print(file_name)
output_pathos = output_pathos+"/"+file_name
if not os.path.exists(output_pathos):
    os.makedirs(output_pathos)
    os.makedirs(output_pathos+"/mask")
    os.makedirs(output_pathos+"/process")
```

代码具体说明如下：

- file_name = os.path.basename(input_pathos)：从输入路径中获取文件名，包括扩展名。
- file_name = file_name.split('.')[0]：将文件名按照扩展名进行分割，只保留文件名部分（去除扩展名）。
- print(file_name)：打印文件名，这一步是为了在控制台显示文件名，供用户查看。
- output_pathos = output_pathos+"/"+file_name：将输出路径设置为输出文件夹路径和文件名的组合。
- if not os.path.exists(output_pathos)：检查输出文件夹是否存在，如果不存在则执行以下操作：
  ◆ os.makedirs(output_pathos)：创建输出文件夹。
  ◆ os.makedirs(output_pathos+"/mask")：在输出文件夹中创建一个名为"mask"的子文件夹。
  ◆ os.makedirs(output_pathos+"/process")：在输出文件夹中创建一个名为"process"的子文件夹。

（3）执行 ETF 滤波器的操作，生成 ETF 滤波后的图像，并保存输入图像的灰度版本供后续使用。具体实现代码如下：

```
time_start=time.time()
ETF_filter = ETF(input_pathos=input_pathos, output_pathos=output_pathos+'/mask',\
     dir_num=direction, kernel_radius=kernel_radius, iter_time=iter_time, background_dir=background_dir)
ETF_filter.forward()
print('ETF done')

input_img = cv2.imread(input_pathos, cv2.IMREAD_GRAYSCALE)
(h0,w0) = input_img.shape
cv2.imwrite(output_pathos + "/input_gray.jpg", input_img)
```

上述代码的主要功能是进行图像处理中的 ETF（edge tangent flow）操作，以及保存输入图像的灰度版本。具体说明如下：

- time_start=time.time()：记录开始时间，用于计算 ETF 操作的执行时间。
- ETF_filter = ETF(input_pathos=input_pathos, output_pathos=output_pathos+'/mask',\ dir_num=direction, kernel_radius=kernel_radius, iter_time=iter_time, background_dir=background_dir)：创建一个名为 ETF_filter 的 ETF 滤波器对象，其中传入的参数包括输入路径、输出路径（包括子文件夹"mask"）、方向数（dir_num）、核半径（kernel_radius）、迭代次数（iter_time）和背景方向（background_dir）。
- ETF_filter.forward()：执行 ETF 滤波器的前向操作，对输入图像进行 ETF 处理。
- print('ETF done')：在控制台打印"ETF done"，表示 ETF 操作已完成。
- input_img = cv2.imread(input_pathos, cv2.IMREAD_GRAYSCALE)：使用 OpenCV 读取输入图像的灰度版本，以便后续处理。
- (h0,w0) = input_img.shape：获取灰度图像的高度和宽度。
- cv2.imwrite(output_pathos + "/input_gray.jpg", input_img)：将灰度图像保存到输出文件夹中，命名为"input_gray.jpg"。

（4）对输入图像进行色调转换（如果 transTone 为 True），创建绘制图像所需的初始变量和数据结构，并将输入图像按不同的角度进行旋转，以准备进行后续的绘制操作。具体实现代码如下：

```
if transTone == True:
    input_img = transferTone(input_img)

now_ = np.uint8(np.ones((h0,w0)))*255
step = 0
if draw_new==True:
    time_start=time.time()
    stroke_sequence=[]
    stroke_temp={'angle':None, 'grayscale':None, 'row':None, 'begin':None, 'end':None}
    for dirs in range(direction):
        angle = -90+dirs*180/direction
        print('angle:', angle)
        stroke_temp['angle'] = angle
        img,_ = rotate(input_img, -angle)
```

对上述代码的具体说明如下：

- if transTone == True：如果变量 transTone 的值为 True，则执行下面的代码块。

- input_img = transferTone(input_img)：调用 transferTone 函数，对 input_img 进行色调转换。
- now_ = np.uint8(np.ones((h0,w0)))*255：创建一个大小为(h0, w0)的二维数组 now_，并用值 255 填充。这个数组将用于绘制图像的笔画信息。
- step = 0：设置变量 step 的初始值为 0。
- if draw_new==True：如果变量 draw_new 的值为 True，则执行下面的代码块。
- time_start=time.time()：记录开始时间，用于计算绘制图像的执行时间。
- stroke_sequence=[]：创建一个空列表 stroke_sequence，用于存储笔画序列信息。
- stroke_temp={'angle':None, 'grayscale':None, 'row':None, 'begin':None, 'end':None}：创建一个字典 stroke_temp，用于存储每个笔画的角度、灰度、行号、起始点和终止点。
- for dirs in range(direction)：对于 0 到 direction-1 的每个值，执行循环。
- angle = -90+dirs*180/direction：计算当前方向的角度值。
- print('angle:', angle)：打印当前角度值。
- stroke_temp['angle'] = angle：将当前角度值存储到 stroke_temp 字典中的'angle'键。
- img,_ = rotate(input_img, -angle)：调用 rotate 函数，对 input_img 进行逆时针旋转-angle 度，并将旋转后的图像存储到 img 变量中。

（5）进行直方图均衡化操作，这样可以提高图像的对比度和亮度分布。如果 CLAHE 变量为 True，则对旋转后的图像 img 进行直方图均衡化，并打印一条消息表示操作已完成。具体实现代码如下：

```
if CLAHE==True:
    img = HistogramEqualization(img)
    print('HistogramEqualization done')
```

（6）进行梯度计算和归一化操作，具体实现代码如下：

```
img_pad = cv2.copyMakeBorder(img, 2*linewidth, 2*linewidth, 2*linewidth, 2*linewidth, cv2.BORDER_REPLICATE)
img_normal = cv2.normalize(img_pad.astype("float32"), None, 0.0, 1.0, cv2.NORM_MINMAX)

x_der = cv2.Sobel(img_normal, cv2.CV_32FC1, 1, 0, ksize=5)
y_der = cv2.Sobel(img_normal, cv2.CV_32FC1, 0, 1, ksize=5)

x_der = torch.from_numpy(x_der) + 1e-12
y_der = torch.from_numpy(y_der) + 1e-12

gradient_magnitude = torch.sqrt(x_der**2.0 + y_der**2.0)
gradient_norm = gradient_magnitude/gradient_magnitude.max()
```

对上述代码的具体说明如下：
- img_pad = cv2.copyMakeBorder(img, 2*linewidth, 2*linewidth, 2*linewidth, 2*linewidth, cv2.BORDER_REPLICATE)：将图像 img 进行边界填充，边界宽度为 2*linewidth，填充方式为复制边界像素值。
- img_normal = cv2.normalize(img_pad.astype("float32"), None, 0.0, 1.0, cv2.NORM_MINMAX)：将填充后的图像 img_pad 转换为浮点型，并进行归一化处理，像素值范围从原始值域映射到[0.0, 1.0]。
- x_der = cv2.Sobel(img_normal, cv2.CV_32FC1, 1, 0, ksize=5)：对归一化后的图像

img_normal 在 x 方向上应用 Sobel 算子，计算 x 方向上的梯度。
- y_der = cv2.Sobel(img_normal, cv2.CV_32FC1, 0, 1, ksize=5)：对归一化后的图像 img_normal 在 y 方向上应用 Sobel 算子，计算 y 方向上的梯度。
- x_der = torch.from_numpy(x_der) + 1e-12 和 y_der = torch.from_numpy(y_der) + 1e-12：将计算得到的 x 和 y 方向上的梯度转换为 PyTorch 张量，并加上一个小的常数 1e-12，用于避免除零错误。
- gradient_magnitude = torch.sqrt(x_der**2.0 + y_der**2.0)：计算梯度幅值，即 x 和 y 方向上梯度的欧氏距离。
- gradient_norm = gradient_magnitude/gradient_magnitude.max()：将梯度幅值进行归一化，除以最大幅值，使得梯度的范围在[0, 1]之间。

（7）进行图像的量化操作，具体实现代码如下：

```
ldr = LDR(img, n)
cv2.imshow('Quantization', ldr)
cv2.waitKey(0)
cv2.imwrite(output_pathos + "/Quantization.png", ldr)
```

对上述代码的具体说明如下：
- ldr = LDR(img, n)：调用函数 LDR 对梯度归一化图像 img 进行量化操作，将图像分为 n 个不同的灰度级别。
- cv2.imwrite(output_pathos + "/Quantization.png", ldr)：将量化后的图像 ldr 保存为文件，文件路径为 output_pathos + "/Quantization.png"，即输出路径下的"Quantization.png"文件。
- cv2.imshow('Quantization', ldr) 和 cv2.waitKey(0)：显示量化图像。

（8）使用 for 循环在每个灰度级别下生成笔画序列，首先设置当前迭代的灰度级别 j*256/n；然后读取对应灰度级别的掩膜图像，并进行归一化处理。读取方向掩膜图像，并对其进行旋转和二值化处理；然后生成高斯分布，用于计算笔画的长度；接下来初始化笔画的起始行位置，遍历高斯分布中的每个值，表示笔画的长度。具体实现代码如下：

```
            LDR_single_add(ldr,n,output_pathos)
            print('Quantization done')


            # get tone
            (h,w) = ldr.shape
            canvas = Gassian((h+4*linewidth,w+4*linewidth), mean=250, var = 3)


            for j in range(n):
                # print('tone:',j)
                # distribution = ChooseDistribution(linewidth=linewidth,Grayscale=j*256/n)
                stroke_temp['grayscale'] = j*256/n
                mask = cv2.imread(output_pathos + '/mask/mask{}.png'.format(j),cv2.IMREAD_ GRAYSCALE)/255
                dir_mask = cv2.imread(output_pathos + '/mask/dir_mask{}.png'.format(dirs),cv2.IMREAD_GRAYSCALE)
                # if angle==0:
                #     dir_mask[::] = 255
                dir_mask,_ = rotate(dir_mask, -angle, pad_color=0)
```

```
                dir_mask[dir_mask<128]=0
                dir_mask[dir_mask>127]=1

                distensce = Gassian((1,int(h/linewidth)+4), mean = linewidth, var = 1)
                distensce  =  np.uint8(np.round(np.clip(distensce,  linewidth*0.8,
linewidth*1.25)))
                raw = -int(linewidth/2)

                for i in np.squeeze(distensce).tolist():
                    if raw < h:
                        y = raw + 2*linewidth # y < h+2*linewidth
                        raw += i
                        for interval in get_start_end(mask[y-2*linewidth]*dir_mask
[y-2*linewidth]):

                            begin = interval[0]
                            end = interval[1]

                            # length = end - begin

                            begin -= 2*linewidth
                            end += 2*linewidth

                            length = end - begin
                            stroke_temp['begin'] = begin
                            stroke_temp['end'] = end
                            stroke_temp['row'] = y-int(linewidth/2)
                            print(gradient_norm[y,interval[0]+2*linewidth:interval[1]
+2*linewidth])

                            stroke_temp['importance'] = (2510-stroke_temp['grayscale'])
*torch.sum(gradient_norm[y:y+linewidth,interval[0]+2*linewidth:interval[1]+2*linewidt
h]).numpy()

                            stroke_sequence.append(stroke_temp.copy())
```

- LDR_single_add(ldr,n,output_pathos)：调用函数 LDR_single_add 将每个灰度级别的量化图像进行累积操作，将结果保存到输出路径中。
- print('Quantization done')：输出提示信息，表示量化操作已完成。
- canvas = Gassian((h+4*linewidth,w+4*linewidth), mean=250, var = 3)：创建一个高斯噪声背景图像，尺寸为(h+4×linewidth,w+4×linewidth)，均值为250，方差为3。
- for j in range(n):：遍历每个灰度级别。

（9）据笔画序列逐步绘制图像，生成最终的绘画结果。在绘制过程中，可以选择显示每个步骤的图像，以及保存中间过程的图像帧用于生成绘画动画。具体实现代码如下：

```
        time_end=time.time()
        print('total time',time_end-time_start)
        print('stoke number',len(stroke_sequence))
        # cv2.imwrite(output_pathos + "/draw.png", now_)
        # cv2.imshow('draw', now_)
        # cv2.waitKey(0)

        if random_order == True:
            random.shuffle(stroke_sequence)

        if ETF_order == True:
            random.shuffle(stroke_sequence)
            quickSort(stroke_sequence,0,len(stroke_sequence)-1)
```

```python
            result = Gassian((h0,w0), mean=250, var = 3)
            canvases = []
            
            for dirs in range(direction):
                angle = -90+dirs*180/direction
                canvas,_ = rotate(result, -angle)
                # (h,w) = canvas.shape
                canvas = np.pad(canvas, pad_width=2*linewidth, mode='constant', constant_values=(255,255))
                canvases.append(canvas)
            
            for stroke_temp in stroke_sequence:
                angle = stroke_temp['angle']
                dirs = int((angle+90)*direction/180)
                grayscale = stroke_temp['grayscale']
                distribution = ChooseDistribution(linewidth=linewidth,Grayscale=grayscale)
                row = stroke_temp['row']
                begin = stroke_temp['begin']
                end = stroke_temp['end']
                length = end - begin
                
                newline = Getline(distribution=distribution, length=length)
                
                canvas = canvases[dirs]
                
                if length<1000 or begin == -2*linewidth or end == w-1+2*linewidth:
                    temp = canvas[row:row+2*linewidth,2*linewidth+begin:2*linewidth+end]
                    m = np.minimum(temp, newline[:,:temp.shape[1]])
                    canvas[row:row+2*linewidth,2*linewidth+begin:2*linewidth+end] = m
                # else:
                #     temp = canvas[row:row+2*linewidth,2*linewidth+begin-2*linewidth:2*linewidth+end+2*linewidth]
                #     m = np.minimum(temp, newline)
                #         canvas[row:row+2*linewidth,2*linewidth+begin-2*linewidth:2*linewidth+end+2*linewidth] = m
                
                now,_ = rotate(canvas[2*linewidth:-2*linewidth,2*linewidth:-2*linewidth], angle)
                (H,W) = now.shape
                now = now[int((H-h0)/2):int((H-h0)/2)+h0, int((W-w0)/2):int((W-w0)/2)+w0]
                result = np.minimum(now,result)
                if process_visible == True:
                    cv2.imshow('step', result)
                    cv2.waitKey(1)
                
                step += 1
                if step % Freq == 0:
                    cv2.imwrite(output_pathos + "/process/{0:04d}.jpg".format(int(step/Freq)),result)
            if step % Freq != 0:
                step = int(step/Freq)+1
                cv2.imwrite(output_pathos + "/process/{0:04d}.jpg".format(step), result)
            
            cv2.destroyAllWindows()
            time_end=time.time()
            print('total time',time_end-time_start)
            print('stoke number',len(stroke_sequence))
```

在上述代码中，stroke_sequence 是一个存储笔画信息的列表。在绘制图像的过程中，通过分析图像的边缘和灰度信息，将图像分解为一系列的笔画（strokes）。每个笔画由以下信息组成：
- angle：笔画的角度（方向）。
- grayscale：笔画的灰度值（亮度）。
- row：笔画的起始行位置。
- begin：笔画的起始列位置。
- end：笔画的结束列位置。
- importance：笔画的重要性（根据灰度值和边缘信息计算得出）。

列表 stroke_sequence 按照一定的规则存储了所有的笔画信息。在绘制图像时，会根据笔画的重要性和其他属性来确定绘制的顺序和方式。快速排序函数 quickSort()用于根据笔画的重要性对列表 stroke_sequence 进行排序，以便在绘制过程中先绘制重要性较高的笔画，从而实现更精确的绘制效果。

（10）将生成的边缘图像与之前的绘画结果图像进行合并，生成最终的结果图像。然后对结果图像进行颜色处理，去除蓝色成分，并将结果图像的亮度通道替换为合成的图像，从而生成最终的彩色图像。具体实现代码如下：

```python
edge = genStroke(input_img,18)
edge = np.power(edge, deepen)
edge = np.uint8(edge*255)
if edge_CLAHE==True:
    edge = HistogramEqualization(edge)

cv2.imwrite(output_pathos + '/edge.jpg', edge)
cv2.imshow("edge",edge)


############# merge #############
edge = np.float32(edge)
now_ = cv2.imread(output_pathos + "/draw.jpg", cv2.IMREAD_GRAYSCALE)
result = res_cross= np.float32(now_)

result[1:,1:] = np.uint8(edge[:-1,:-1] * res_cross[1:,1:]/255)
result[0] = np.uint8(edge[0] * res_cross[0]/255)
result[:,0] = np.uint8(edge[:,0] * res_cross[:,0]/255)
result = edge*res_cross/255
result=np.uint8(result)

cv2.imwrite(output_pathos + '/result.jpg', result)
# cv2.imwrite(output_pathos + "/process/{0:04d}.png".format(step+1), result)
cv2.imshow("result",result)


# dellblue
dellblue(result, output_pathos)

# RGB
img_rgb_original = cv2.imread(input_pathos, cv2.IMREAD_COLOR)
cv2.imwrite(output_pathos + "/input.jpg", img_rgb_original)
img_yuv = cv2.cvtColor(img_rgb_original, cv2.COLOR_BGR2YUV)
img_yuv[:,:,0] = result
img_rgb = cv2.cvtColor(img_yuv, cv2.COLOR_YUV2BGR)

cv2.imshow("RGB",img_rgb)
cv2.waitKey(0)
cv2.imwrite(output_pathos + "/result_RGB.jpg",img_rgb)
```

对上述代码的具体说明如下：
- 生成边缘图像：使用 genStroke 函数生成输入图像的边缘图像，将其深化并转换为灰度图像。
- 如果需要，对边缘图像进行直方图均衡化。
- 将边缘图像保存为文件，并显示边缘图像。
- 将边缘图像和之前生成的绘画结果图像进行合并，生成最终的结果图像。
- 将结果图像保存为文件，并显示结果图像。
- 进行去蓝色处理，删除结果图像中的蓝色成分。
- 读取输入图像的彩色原始图像，并将结果图像的亮度通道替换为结果图像。
- 将结果图像转换回彩色图像，并保存为文件。

代码执行后的效果如图 9-3 所示。

图 9-3　代码执行后绘制的效果

9.5.2　绘制铅笔画

编写文件 process_order.py，功能是使用一系列的图像处理函数和方法绘制指定的铅笔画，包含的图像处理方法有 ETF 滤波、直方图均衡化、梯度计算、量化、绘制线条等。和前面的文件 cat.py 相比，保存绘图结果的路径不同：文件 process_order.py 将处理结果保存在指定的路径下的不同文件夹中，而文件 cat.py 直接显示结果图像，不保存文件。文件 process_order.py 的具体实现流程如下：

（1）导入所需的库和模块。
（2）设置程序运行所需的参数。
（3）创建输出文件夹和子文件夹。
（4）进行边缘流动场（ETF）滤波。
（5）读取输入图像，并进行一些预处理操作。
（6）根据给定的方向数量，生成一系列笔画序列。
（7）根据笔画序列，将笔画逐步添加到画布上，形成最终的效果图。
（8）生成边缘图像。
（9）将边缘图像与最终效果图进行合并。
（10）进行图像后处理操作，如去除蓝色。
（11）将结果转换为 RGB 格式。
（12）保存生成的结果图像。

该代码通过将图像进行量化、边缘检测和笔画生成等步骤，实现将输入图像转换为类似铅笔画效果的输出图像。文件 process_order.py 的主要实现代码如下：

```
edge = genStroke(input_img,18)
edge = np.power(edge, deepen)
edge = np.uint8(edge*255)
if edge_CLAHE==True:
    edge = HistogramEqualization(edge)
```

```
cv2.imwrite(output_pathos + '/edge.jpg', edge)
cv2.imshow("edge",edge)


############# merge #############
edge = np.float32(edge)
now_ = cv2.imread(output_pathos + "/draw.jpg", cv2.IMREAD_GRAYSCALE)
result = res_cross= np.float32(now_)

result[1:,1:] = np.uint8(edge[:-1,:-1] * res_cross[1:,1:]/255)
result[0] = np.uint8(edge[0] * res_cross[0]/255)
result[:,0] = np.uint8(edge[:,0] * res_cross[:,0]/255)
result = edge*res_cross/255
result=np.uint8(result)

cv2.imwrite(output_pathos + '/result.jpg', result)
# cv2.imwrite(output_pathos + "/process/{0:04d}.png".format(step+1), result)
cv2.imshow("result",result)


# dellblue
dellblue(result, output_pathos)

# RGB
img_rgb_original = cv2.imread(input_pathos, cv2.IMREAD_COLOR)
cv2.imwrite(output_pathos + "/input.jpg", img_rgb_original)
img_yuv = cv2.cvtColor(img_rgb_original, cv2.COLOR_BGR2YUV)
img_yuv[:,:,0] = result
img_rgb = cv2.cvtColor(img_yuv, cv2.COLOR_YUV2BGR)

cv2.imshow("RGB",img_rgb)
cv2.waitKey(0)
cv2.imwrite(output_pathos + "/result_RGB.jpg",img_rgb)
```

代码执行后的效果如图 9-4 所示。

图 9-4　代码执行后绘制的铅笔画效果

第 10 章 小区 AI 停车计费管理系统开发

在国内经济飞速发展的今天，小区停车难、高峰期进出小区困难的问题日益严重。通过使用 AI 技术自动识别车牌，并实现实时计费功能可以提高车辆的通行效率。本章详细介绍使用百度 AI 技术开发一个小区停车计费管理系统的过程。

10.1 背景介绍

随着我国汽车保有量的快速增长，除了停车难和停车场进出拥堵之外，高昂的值班人员人工成本问题，收费跑、漏等现象也经常发生。

信息技术的快速进步，让通过较低的投入实现降本增收成为现实，通过智能化的系统方案，让小区出入口实现无人值守化管理。针对临时停车和有固定车位的用户，推出车牌识别停车计费管理系统方案，利用车牌识别技术取代传统的 IC 卡技术，解决车辆进出时必须停下刷卡而造成的停车场进出口堵车现象，这种智能的停车管理系统为停车用户提供了一种崭新的服务模式。通过该系统，有固定停车位的车辆进出可以实现不停车通行，临时车辆入场不停车出场缴费自动放行，整个系统结构简单，稳定可靠，安装、维护、使用方便。

10.2 系统功能分析和模块设计

这个小区停车计费管理系统是基于百度人工智能技术实现的，节省了物业的人员成本，提高了小区车辆的通行效率，为大家的出行带来了极大的方便。

10.2.1 功能分析

区别于传统的 IC 卡、蓝牙卡停车计费模式，本系统采用百度 AI 技术识别车牌信息，通过车牌号码判别是临时还是固定车辆，从而对进出车辆进行计费管理。应用文通车牌识别停车计费系统具有以下特点：

（1）针对固定车辆进场出场都无须刷卡，车牌识别后自动放行，车辆进出效率明显提升。

（2）针对临时车辆进场无须停车取卡，进场车牌识别后道闸自动放行，出场时车牌识别停车缴费后，道闸自动放行。

（3）车牌识别停车管理模式免去了物业管理 IC 卡、蓝牙卡的成本，后续维护成本低。

（4）无论是固定还是临时车辆，车主都无须为卡未带、卡丢失、卡损坏而发愁，车主用户体验大大提升。

（5）由于采用车牌识别停车管理模式后，车辆进出效率大大提高，物业停车收益也会明显提升。

10.2.2 系统模块设计

本项目的功能模块如图 10-1 所示。

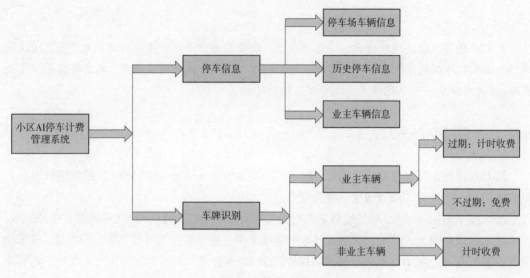

图 10-1 功能模块

10.3 系统 GUI

本系统是使用库 Pygame 实现的桌面项目,通过可视化桌面识别进出小区的车辆信息,并自动实现计费功能。

10.3.1 设置基本信息

编写文件 settings.py,功能是设置 Pygame 桌面的基本信息,包括屏幕大小、颜色和停车位总数等信息。主要实现代码如下:

```
class Settings():
    def __init__(self):
        """ 初始化设置 """
        # 屏幕设置(宽、高、背景色、线颜色)
        self.screen_width = 1000
        self.screen_height = 484
        self.bg_color = (255, 255, 255)

        # 停车位
        self.total = 100

        # 识别颜色、车牌号、进来时间、出入场信息
        self.ocr_color = (212, 35, 122)
        self.carnumber = ''
        self.comeInTime = ''
        self.message = ''
```

10.3.2 绘制操作按钮

编写文件 button.py，功能是在 GUI 界面绘制"识别"按钮，具体实现代码如下：

```python
import pygame.font
class Button():
    def __init__(self, screen, msg):
        """初始化按钮的属性"""
        self.screen = screen
        self.screen_rect = screen.get_rect()
        # 设置按钮的尺寸和其他属性
        self.width, self.height = 100, 50
        self.button_color = (0, 120, 215)
        self.text_color = (255, 255, 255)
        self.font = pygame.font.SysFont('SimHei', 25)
        # 创建按钮的 rect 对象，并使其居中
        self.rect = pygame.Rect(0, 0, self.width, self.height)
        # 创建按钮的 rect 对象，并设置按钮中心位置
        self.rect.centerx = 640 - self.width / 2 + 2
        self.rect.centery = 480 - self.height / 2 + 2
        # 按钮的标签只需创建一次
        self.prep_msg(msg)
    def prep_msg(self, msg):
        """将 msg 渲染为图像，并使其在按钮上居中"""
        self.msg_image = self.font.render(msg, True, self.text_color, self.button_color)
        self.msg_image_rect = self.msg_image.get_rect()
        self.msg_image_rect.center = self.rect.center
    def draw_button(self):
        # 绘制一个用颜色填充的按钮，再绘制文本
        self.screen.fill(self.button_color, self.rect)
        self.screen.blit(self.msg_image, self.msg_image_rect)
```

10.3.3 绘制背景和文字

编写文件 textboard.py，功能是绘制 GUI 的背景和文字，分别设置背景图案、绘制线条和识别结果文字的属性。具体实现代码如下：

```python
import pygame.font

# 线颜色
line_color = (0, 0, 0)
# 显示文字信息时使用的字体设置
text_color = (0, 0, 0)

def draw_bg(screen):
    # 背景文图案
    bgfont = pygame.font.SysFont('SimHei', 15)
    # 绘制横线
    pygame.draw.aaline(screen, line_color, (662, 30), (980, 30), 1)
    # 渲染为图片
    text_image = bgfont.render('识别信息: ', True, text_color)
    # 获取文字图像位置
    text_rect = text_image.get_rect()
    # 设置文字图像中心点
```

```
        text_rect.left = 660
        text_rect.top = 370
        # 绘制内容
        screen.blit(text_image, text_rect)

# 绘制文字（text 是文字内容、xpos 是 x 坐标、ypos 是 y 坐标、fontSize 是字体大小）
def draw_text(screen, text, xpos, ypos, fontsize, tc=text_color):
    # 使用系统字体
    xtfont = pygame.font.SysFont('SimHei', fontsize)
    text_image = xtfont.render(text, True, tc)
    # 获取文字图像位置
    text_rect = text_image.get_rect()
    # 设置文字图像中心点
    text_rect.left = xpos
    text_rect.top = ypos
    # 绘制内容
    screen.blit(text_image, text_rect)
```

10.4 车牌识别和收费

车牌识别是本系统的核心，为了提高开发效率，本项目使用百度在线 AI 技术识别车牌。

10.4.1 登记业主的车辆信息

在表格文件"业主车辆信息表.xlsx"中登记业主的车辆信息，包括车牌号和到期时间，如图 10-2 所示。

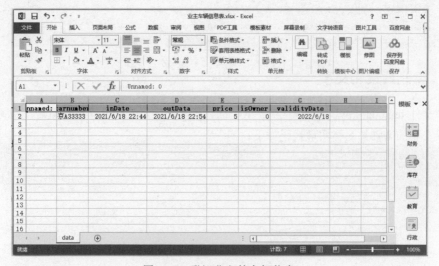

图 10-2　登记业主的车辆信息

10.4.2 识别车牌

编写文件 ocrutil.py，功能是调用百度 AI 中的文字识别 SDK 获取图片中的车牌信息。注意，在测试时图片文件"test.jpg"表示为从摄像头读取的图片，每次循环获取一次，也可以使用 MP4 视频进行测试。具体实现代码如下：

```
from aip import AipOcr
import os

# 百度识别车牌
# 申请地址 https://login.bce.baidu.com/
filename = 'file/key.txt'  # 记录申请的 Key 的文件位置
if os.path.exists(filename):  # 判断文件是否存在
    with open(filename, "r") as file:  # 打开文件
        dictkey = eval(file.readlines()[0])  # 读取全部内容转换为字典
        # 以下获取的三个 Key 是进入百度 AI 开放平台的控制台的应用列表里创建应用得来的
        APP_ID = dictkey['APP_ID']  # 获取申请的 APIID
        API_KEY = dictkey['API_KEY']  # 获取申请的 APIKEY
        SECRET_KEY = dictkey['SECRET_KEY']  # 获取申请的 SECRETKEY
else:
    print("请先在 file 目录下创建 key.txt,并且写入申请的 Key! 格式如下: "
          "\n{'APP_ID':'申请的 APIID', 'API_KEY':'申请的 APIKEY', 'SECRET_KEY':'申请的 SECRETKEY'}")
# 初始化 AipOcr 对象
client = AipOcr(APP_ID, API_KEY, SECRET_KEY)

# 读取文件
def get_file_content(filePath):
    with open(filePath, 'rb') as fp:
        return fp.read()

# 根据图片返回车牌号
def getcn():
    # 读取图片
    image = get_file_content('images/test.jpg')
    # 调用车牌识别
    results = client.licensePlate(image)['words_result']['number']
    # 输出车牌号
    return results
```

10.4.3 计算停车时间

编写文件 timeutil.py,功能是根据是否为业主车辆来计算停车时间,主要包含如下两个功能函数:

- 函数 time_cmp():用来比较出场时间跟卡的有效期,判断业主是否需要收费;
- 函数 priceCalc():用来计算停车时间。存在两种情况,一种是外来车,只需要比较出入场时间差;另一种是业主车,入场时卡未到期,但出场时已经到期,所以需要比较卡有效期和出场时间的差值。

文件 timeutil.py 的具体实现代码如下:

```
import datetime
import math

# 计算两个日期大小
def time_cmp(first_time, second_time):
    # 由于有效期获取后会有小数数据
    firstTime  =  datetime.datetime.strptime(str(first_time).split('.')[0], "%Y-%m-%d %H:%M:%S")
    secondTime = datetime.datetime.strptime(str(second_time), "%Y-%m-%d %H:%M")
```

```
            number = 1 if firstTime > secondTime else 0
            return number

    # 计算停车时间
    def priceCalc(inDate, outDate):
        if '.' in str(inDate):
            inDate = str(inDate).split('.')[0]
            inDate = datetime.datetime.strptime(inDate, "%Y-%m-%d %H:%M:%S")
            print('特殊处理')
        else:
            inDate = datetime.datetime.strptime(inDate, "%Y-%m-%d %H:%M")
        outDate = datetime.datetime.strptime(str(outDate), "%Y-%m-%d %H:%M")
        rtn = outDate - inDate
        # 计算停车多少小时（往上取整）
        y = math.ceil(rtn.total_seconds() / 60 / 60)
        return y
```

注意：在上述代码中，由于要读取 Excel 的卡有效期字段，会多出".xxxx"这部分，所以需要经过 split('.')处理。

10.4.4 识别车牌并计费

编写文件 procedure_functions.py，当单击"识别"按钮后识别车牌并计算停车费，主要分为如下两种处理逻辑：

第一种：当停车场未有停车时，只需要识别后，把车辆信息存入"停车场车辆表"并把相关信息显示到界面右下角。

第二种：当停车场已有停车时，会出现入场和出场两种情况。

在车辆入场时需判断是否停车场已满，已满则不给进入并显示提示信息。如果未满则需要把车辆信息存入"停车场车辆表"并把相关信息显示到界面右下角。

在车辆出场时分业主有效、业主过期和外来车三种情况收费，删除车辆表相应的车辆信息，并把车辆信息和收费信息等存入"停车场历史表"（可用于后面数据的汇总统计）。

文件 procedure_functions.py 的具体实现代码如下：

```
import sys
import pygame
import time
import pandas as pd
import ocrutil
import timeutil

# 事件
def check_events(settings, recognition_button, ownerInfo_table, carInfo_table, history_table, path):
    """ 响应按键和鼠标事件 """
    for event in pygame.event.get():
        if event.type == pygame.QUIT:
            sys.exit()
        elif event.type == pygame.MOUSEBUTTONDOWN:
            mouse_x, mouse_y = pygame.mouse.get_pos()
            button_clicked = recognition_button.rect.collidepoint(mouse_x, mouse_y)
            if button_clicked:
                try:
```

```python
                # 获取车牌
                carnumber = ocrutil.getcn()

                # 转换当前时间的格式，格式例如：2022-12-11 16:18
                localtime = time.strftime('%Y-%m-%d %H:%M', time.localtime())
                settings.carnumber = '车牌号码：' + carnumber

                # 判断进入车辆是否是业主车辆
                # 获取业主车辆信息（只显示卡未过期）
                ownerInfo_table = ownerInfo_table[ownerInfo_table['validityDate'] > localtime]
                owner_carnumbers = ownerInfo_table[['carnumber', 'validityDate']].values
                carnumbers = ownerInfo_table['carnumber'].values
                # 获取车辆表信息
                carInfo_carnumbers = carInfo_table[['carnumber', 'inDate', 'isOwner', 'validityDate']].values
                cars = carInfo_table['carnumber'].values
                # 增加车辆信息
                append_carInfo = {
                    'carnumber': carnumber
                }
                # 增加历史信息
                append_history = {
                    'carnumber': carnumber
                }
                carInfo_length = len(carInfo_carnumbers)
                # 车辆表未有数据
                if carInfo_length == 0:
                    print('目前车辆进入小区')
                    in_park(owner_carnumbers, carnumbers, carInfo_table, append_carInfo, carnumber, localtime,
                            settings, path)
                # 车辆表有数据
                else:
                    if carnumber in cars:
                        # 出停车场
                        i = 0
                        for carInfo_carnumber in carInfo_carnumbers:
                            if carnumber == carInfo_carnumber[0]:
                                if carInfo_carnumber[2] == 1:
                                    if timeutil.time_cmp(carInfo_carnumber[3], localtime):
                                        print('业主车，自动抬杆')
                                        msgMessage = '业主车，可出停车场'
                                        parkPrice = '业主卡'
                                    else:
                                        print('这是业主车，但是已经已过期，需要收费抬杆')
                                        # 比较卡有效期时间
                                        price = timeutil.priceCalc(carInfo_carnumber[3], localtime)
                                        msgMessage = '停车费用：' + str(5 * int(price)) + '(业主您好，您的卡已到期)'
                                        parkPrice = 5 * int(price)
                                else:
                                    print('外来车，收费抬杆')
```

```python
                                # 比较入场时间
                                price = timeutil.priceCalc(carInfo_carnumber[1], localtime)
                                msgMessage = '停车费用: ' + str(5 * price)
                                parkPrice = 5 * int(price)

                            print(i)
                            carInfo_table = carInfo_table.drop([i])
                            # 增加数据到历史表
                            append_history['inDate'] = carInfo_carnumber[1]
                            append_history['outData'] = localtime
                            append_history['price'] = parkPrice
                            append_history['isOwner'] = carInfo_carnumber[2]
                            append_history['validityDate'] = carInfo_carnumber[3]
                            history_table = history_table.append(append_history, ignore_index=True)

                            settings.comeInTime = '出场时间: ' + localtime
                            settings.message = msgMessage

                            # 更新车辆表和历史表
                            pd.DataFrame(carInfo_table).to_excel(path + '停车场车辆表' + '.xlsx', sheet_name='data',
                                                                  index=False, header=True)
                            pd.DataFrame(history_table).to_excel(path + '停车场历史表' + '.xlsx', sheet_name='data',
                                                                  index=False, header=True)
                            break
                        i += 1
                else:
                    # 入停车场
                    print('有车辆表数据入场')
                    if carInfo_length < settings.total:
                        in_park(owner_carnumbers, carnumbers, carInfo_table, append_carInfo, carnumber,
                                localtime, settings, path)
                    else:
                        print('停车场已满')
                        settings.comeInTime = '进场时间: ' + localtime
                        settings.message = '停车场已满，无法进入'
            except Exception as e:
                print("错误原因: ", e)
                continue
            pass


# 车辆入停车场
def in_park(owner_carnumbers, carnumbers, carInfo_table, append_carInfo, carnumber, localtime, settings, path):
    if carnumber in carnumbers:
        for owner_carnumber in owner_carnumbers:
            if carnumber == owner_carnumber[0]:
                print('业主车，自动抬杆')
                msgMessage = '提示信息: 业主车，可入停车场'
```

```python
                append_carInfo['isOwner'] = 1
                append_carInfo['validityDate'] = owner_carnumber[1]
                # 退出循环
                break
        else:
            print('外来车,识别抬杆')
            msgMessage = '提示信息: 外来车,可入停车场'
            append_carInfo['isOwner'] = 0
    append_carInfo['inDate'] = localtime
    settings.comeInTime = '进场时间: ' + localtime
    settings.message = msgMessage
    # 添加信息到车辆表
    carInfo_table = carInfo_table.append(append_carInfo, ignore_index=True)
    # 更新车辆表
    pd.DataFrame(carInfo_table).to_excel(path + '停车场车辆表' + '.xlsx', sheet_name='data',
                                    index=False, header=True)
```

10.5 主程序

本项目的主程序文件是 main.py,功能是编写主函数初始化程序,并调用上面的功能函数展示 GUI 桌面,监听用户单击按钮实现车辆识别和计费处理。文件 main.py 的具体实现代码如下:

```python
import pygame
import cv2
import os
import pandas as pd
# 引入自定义模块
from settings import Settings
from button import Button
import textboard
import procedure_functions as pf

def run_procedure():
    # 获取文件的路径
    cdir = os.getcwd()
    # 文件夹路径
    path = cdir + '/file/'
    # 读取路径
    if not os.path.exists(path + '停车场车辆表' + '.xlsx'):
        # 车牌号 进入时间 离开时间 价格 是否业主
        carnfile = pd.DataFrame(columns=['carnumber', 'inDate', 'outData', 'price', 'isOwner', 'validityDate'])
        # 生成xlsx文件
        carnfile.to_excel(path + '停车场车辆表' + '.xlsx', sheet_name='data')
        carnfile.to_excel(path + '停车场历史表' + '.xlsx', sheet_name='data')

    settings = Settings()
    # 初始化并创建一个屏幕对象
    pygame.init()
    pygame.display.set_caption('小区智能车牌识别系统')
    ic_launcher = pygame.image.load('images/icon_launcher.png')
    pygame.display.set_icon(ic_launcher)
    screen = pygame.display.set_mode((settings.screen_width, settings.screen_height))
```

```python
try:
    # cam = cv2.VideoCapture(0)  # 开启摄像头
    cam = cv2.VideoCapture('file/test.mp4')
except:
    print('请连接摄像头')
# 循环帧率设置
clock = pygame.time.Clock()
running = True
# 开始主循环
while running:
    screen.fill(settings.bg_color)
    # 从摄像头读取图片
    sucess, img = cam.read()
    # 保存图片,并退出
    if sucess:
        cv2.imwrite('images/test.jpg', img)
    else:
        # 识别不到图片或者设备停止,则退出系统
        running = False
    # 加载图像
    image = pygame.image.load('images/test.jpg')
    # 设置图片大小
    image = pygame.transform.scale(image, (640, 480))
    # 绘制视频画面
    screen.blit(image, (2, 2))

    # 创建识别按钮
    recognition_button = Button(screen, '识别')
    recognition_button.draw_button()
    # 读取文件内容
    ownerInfo_table = pd.read_excel(path + '住户车辆表.xlsx', sheet_name='data')
    carInfo_table = pd.read_excel(path + '停车场车辆表.xlsx', sheet_name='data')
    history_table = pd.read_excel(path + '停车场历史表.xlsx', sheet_name='data')
    inNumber = len(carInfo_table['carnumber'].values)
    # 绘制背景
    textboard.draw_bg(screen)
    # 绘制信息标题
    textboard.draw_text(screen, '共有车位: ' + str(settings.total) + '  剩余车位: ' + str(settings.total - inNumber), 680, 0, 20)
    # 绘制信息表头
    textboard.draw_text(screen, '  车牌号         进入时间', 700, 40, 15)
    # 绘制停车场车辆前十条信息
    carInfos = carInfo_table.sort_values(by='inDate', ascending=False)
    i = 0
    for carInfo in carInfos.values:
        if i >= 10:
            break
        i += 1
        textboard.draw_text(screen, str(carInfo[1])+'   '+str(carInfo[2]), 700, 40 + i * 30, 15)
    # 绘制识别信息
    textboard.draw_text(screen, settings.carnumber, 660, 400, 15, settings.ocr_color)
    textboard.draw_text(screen, settings.comeInTime, 660, 422, 15, settings.ocr_color)
    textboard.draw_text(screen, settings.message, 660, 442, 15, settings.ocr_color)
```

```
        """ 响应鼠标事件 """
        pf.check_events(settings, recognition_button, ownerInfo_table, carInfo_table,
history_table, path)
        pygame.display.flip()
        # 控制游戏最大帧率为 60
        clock.tick(60)
    # 关闭摄像头
    cam.release()
run_procedure()
```

执行后单击"识别"按钮可以识别出当前进入小区的车辆信息,如图 10-3 所示。

图 10-3 车辆进小区

当车辆出小区时可以实时识别出车辆信息,并计算出停车费用,如图 10-4 所示。如果是有效期内的业主车辆则免费。

图 10-4 车辆出小区计费

第 11 章　机器人智能物体识别系统开发

随着科技进步和发展，物体识别技术被广泛应用于人们的生产生活中。近年来，随着深度学习与云计算的跳跃式发展，带动了物体识别技术产生质的飞跃。高分辨率图像和检测的实时性要求越来越高。本章详细讲解使用人工智能技术为移动机器人开发一个物体识别检测系统的过程。

11.1　背景介绍

随着机电一体化技术的快速发展，作为其典型代表的机器人的智能化程度越来越高。而工作在复杂环境中的机器人通过使用视觉技术，对周围环境中的物体进行精确识别是机器人智能化的重要标志。与传统机器人不同，具有"视觉"且能够识别物体的机器人可以对外部世界进行感知（即获取图像），分析所得信息，并做出合理的决策。这种技术恰恰满足了对机器人智能化的需求，对机器人的工作和未来机器人的发展具有重要的意义。

机器人视觉的核心技术在于物体识别。物体识别通俗来说就是运用计算机技术使机器人具有和人类一样的，对于在任意环境下观察到的任意物体进行检测、分割和识别的能力。物体识别的作用非常大：对汽车或车牌的识别，并附以其他处理（速度计算等），可以对交通进行智能监控；工厂中智能机器人可以识别零件种类，以对零件进行相应操作（搬运、组装等）；家用机器人对各种物体的识别可以帮助人类做更多的工作，而不是像传统机器人那样只能做一些简单的重复性的事情，这会使机器人更加智能化，发挥更大的作用。在各种各样的物体识别中，人脸识别是最典型的识别之一。准确来讲应该是人脸检测，两者的区别在于，"识别"（recognition）是从图像中找到能与特定人脸相匹配的部分；而"检测"（detection）只是识别的一部分，即在图像中检测出人脸并标记位置。而人脸检测已经满足"物体识别"的要求，它完全可以代表其他物体（如汽车、杯子等）的识别，且人脸检测可以被应用在很多领域。比如家用机器人可以从复杂环境中判断主人的位置，数码相机可以通过人脸识别来对人脸进行准确对焦等等。

11.2　物体识别

大千世界的物体种类繁多，人们主要通过视觉系统对形形色色的物体进行分类和辨别，统称为物体识别。通过模拟人类视觉系统的视觉信息获取和处理功能以便于计算机具有人类识别物体的能力，出现了计算机视觉和模式识别等研究领域。物体识别（object recognition）是当前国内外计算机视觉与模式识别领域的一个活跃的研究方向，在很多方面有了很大的进步。比如对人类的视觉系统有了更进一步的认识、数学工具更高级、计算效率越来越高、越来越多的具有挑战性的数据库收集等，这些进步使得物体识别越来越引起人们的关注。

11.2.1 物体识别介绍

物体识别是机器智能的基本功能之一，它是任何一个以图像或视频作为输入的实际应用系统中的核心问题和关键技术。这类系统的性能和应用前景都依赖于其中物体的知识表示和分类识别所能达到的水平。物体识别技术无论是在军事还是在民用中都有着广泛需求和应用。如智能视频监控、视觉导航、人机交互、计算机取证、各类身份识别和认证系统、数字图书馆和 Internet 中的在海量图像库和视频中的基于内容的检索、编码与压缩等等。

基于图像的物体识别的过程通常表现为：首先建立待识别物体图像的一种知识表示模型，在一定量的训练样本中学习得到一组满足预定要求的模型参数；同时根据物体图像的表示模型，建立一套从实际图像中进行推理的识别算法，通过在实际图像中测试可获得系统的泛化能力对其进行性能评估。由此可见，物体识别技术的提高，无论在军事还是民用方面都有非常重要的意义。

所谓一般物体识别，通俗来说就是使计算机具有和人类一样的，对于在任意环境下观察到的任意物体进行检测、分割和识别的能力。它作为计算机视觉领域的一个特定而又极为重要的任务，要求在给予一定量的训练样本的前提下，计算机能够学习有关指定物体类别的知识，并在观察到从属于旧类别的新物体时，给出识别的结果。

研究一般物体识别，无论对于理论还是实践都有极其重大的意义。计算机视觉的核心在于识别，而一般物体识别又是识别中最为复杂核心的问题，对于神经科学而言，破解了一般物体识别就相当于将神秘莫测的人脑工作机制拉开了帷幕一角，由此展开此后进一步的深入研究，意义不言而喻。在实践中，一般物体识别的研究则能给人类生活的方方面面，尤其是交通、国防、教育带来极为重大的影响，甚至改变人们生活的方式，对整个社会有着深远的意义。

一般物体识别与特定物体识别（specific object recognition）的主要区别在于：特定物体识别通过构造高度特化的特征提取及机器学习方法，使用海量的训练样本进行训练，仅仅处理某种物体或是某类物体，典型的例子如汽车检测及人脸检测；一般物体识别面对的问题则要困难得多，概括来说，它必须使用物体类间通用的一般特征，而不能为某个特定类别定义特征，它必须能处理多类分类及增量学习，在此前提下无法使用给定类别的海量样本进行训练。

11.2.2 物体识别的挑战

在目前的技术条件下，已有的大多数物体表示与识别方法是针对特定的物体实例和表现形式，如字符、人脸、车和车牌等。这种情况下，建模、学习、推理与数据都有很强的针对性，从而也就缺少了通用性和可扩充性。在识别几百类常见物体时，会导致物体识别在建模、学习、推理与数据四个方面都遇到很大的挑战，具体而言有以下几点：

- 不同姿态和视角：因为物体在图像中出现的姿态和视角是任意变化的，系统也无法预先知道物体的详细姿态和视角，在不同姿态和视角下，对识别能起到作用的特征是大不相同的。
- 光照的影响：在不同的光照条件下，物体的正确识别率将会急剧下降。其主要原因是硬件的成像与光照的关系并不是线性的。当光线太强时会出现饱和问题；当光线较弱时，阴影部分就会出现信噪比较低的问题。因此，对于如何设计一种有效地避免光线影响对物体识别的影响来说，解决光线问题是关键。
- 遮挡问题：这也是物体识别中无法回避的问题之一。在真实图像中，待识别的物体很有可能被其他物体部分遮挡，就会丧失一些非常重要的信息，使准确识别目标物体的难度增大。设计一种在目标物体图像因部分遮挡和部分缺失的情况下也能被正确识别的方法，包括最小特征集的确定和选择、特征信息缺失时的识别方法以及遮挡率与识别有效性之间关系的研究就显得十分重要。
- 尺度变化问题：同一类物体可能在大小方面存在较大的差异。图像的各种结构只存在于一定的尺度范围内。如何解决尺度变化问题也是物体识别所要解决的一大问难。这就给同一类物体的识别带来了很大的困难。在早期的图像处理和表示中就遇到一个很大的困难，即图像的描述是依赖于图像的尺度的。
- 形状变化问题：同一类物体的形状变化也会使得物体识别的难度加大。比如椅子、桌子等物体几乎都有成百上千种形状，这样就无法采用一种统一的表示方法来表示物体，导致识别难度加大。因此，设计一种能有效地针对同一类物体形状发生变化而对物体识别造成的影响，对物体识别来说也是一个巨大的挑战。

11.2.3 图像特征的提取方法

图像特征提取就是提取出一幅图像中不同于其他图像的根本属性，以区别不同的图像。如灰度、亮度、纹理和形状等等特征都是与图像的视觉外观相对应的；还有一些则缺少自然的对应性，如颜色直方图、灰度直方图和空间频谱图等。基于图像特征进行物体识别实际上是根据提取到图像的特征来判断图像中物体属于什么类别。形状、纹理和颜色等特征是最常用的视觉特征，也是现阶段基于图像的物体识别技术中采用的主要特征。下面分别介绍一下图像的形状、纹理和颜色特征的提取方法。

1. 图像形状特征提取

形状特征是反映出图像中物体最直接的视觉特征，大部分物体可以通过分辨其形状来进行判别。所以，在物体识别中，形状特征的正确提取显得非常重要。常用的图像形状特征提取方法有两种：基于轮廓的方法和基于区域的方法。这两种方法的不同之处在于：
- 对于基于轮廓的方法来说，图像的轮廓特征主要针对物体的外边界，描述形状的轮廓特征的方法主要有样条、链码和多边形逼近等。
- 而在基于区域的方法中，图像的区域特征则关系到整个形状区域，描述形状的区域特征的主要方法有区域的面积、凹凸面积、形状的主轴方向、纵横比、形状的不变矩等。

关于形状的特征目前已得到了广泛的应用，典型的形状特征描述方法有：
- 边界特征法：该方法的基本思想是通过描述图像的边界特征来获取相应的图像形状参数。其中，边界的方向直方图方法和 Hough 变换检测平行直线的方法是比较经典的方法。边界方向直方图法首先对图像进行微分以求得图像边缘，然后，做出关于边缘方向和大小的直方图，通常采用构造图像灰度梯度方向矩阵的方法。Hough 变换检测平行直线的方法是利用图像全局特性将边缘像素连接起来并组成区域封闭边界的一种方法，其基本思想是利用点到线之间的对偶性进行检测的。
- 傅里叶形状描述符法：该方法的基本思想是对图像中物体边界点作傅里叶变换以作为形

状描述。傅里叶变换主要是利用区域边界的周期性和封闭性,将二维问题转化为一维问题。采用这种方法就可以由物体的边界点导出质心距离、曲率函数和复坐标函数三种形状特征的表达。
- 几何参数法:几何参数法是一种更为简单的图像区域特征描述方法,例如采用有关形状定量测度(如矩、面积、周长等)的形状参数法。这种定量测度简单并且可操作性强,在简单的三维物体识别中可以采用。
- 形状不变矩法:该方法的主要思想是利用目标所占区域的矩作为形状描述参数。矩特征主要表征了图像区域的几何特征,又称为几何矩,由于其具有旋转、平移、尺度等特性的不变特征,所以又称其为不变矩。在图像处理中,几何不变矩可以作为一个重要的特征来表示物体,可以据此特征来对图像进行分类等操作。

2. 图像纹理特征提取

图像的纹理是与物体表面结构和材质有关的图像的内在特征,反映出来的是图像的全局特征。图像的纹理可以描述为:一个邻域内像素的灰度级发生变化的空间分布规律,包括表面组织结构、与周围环境关系等许多重要的图像信息。典型的图像纹理特征提取方法有:统计方法(灰度共生矩阵纹理特征分析方法就是典型的统计方法之一)、几何法(建立在基本的纹理元素理论基础上的一种纹理特征分析方法)、模型法(将图像的构造模型的参数作为纹理特征)和信号处理法(主要是小波变换为主)。

3. 图像颜色特征提取

图像的颜色特征描述了图像或图像区域的物体的表面性质,反映出的是图像的全局特征。一般来说,图像的颜色特征是基于像素点的特征,只要是属于图像或图像区域内的像素点都将会有贡献。典型的图像颜色特征提取方法有:颜色直方图、颜色集和颜色矩。具体说明如下:
- 颜色直方图:颜色直方图是最常用的表达颜色特征的方法,它的优点是能简单描述图像中不同色彩在整幅图像中所占的比例,特别适用于描述一些不需要考虑物体空间位置的图像和难以自动分割的图像。而颜色直方图的缺点是它无法描述图像中的某一具体的物体,无法区分局部颜色信息。
- 颜色集:颜色集可以看成是颜色直方图的一种近似表达。具体方法是:首先将图像从 RGB 颜色空间转换到视觉均衡的颜色空间;然后将视觉均衡的颜色空间量化;最后,采用色彩分割技术自动将图像分为几个区域,用量化的颜色空间中的某个颜色分量来表示每个区域的索引,这样就可以用一个二进制的颜色索引集来表示一幅图像了。
- 颜色矩:颜色矩方法是基于图像中任何的颜色分布都可以用相应的矩来表示这个数学基础上的。由于颜色分布信息主要集中在低阶矩中,因此,表达图像的颜色分布仅需要采用颜色的一阶矩(mean)、二阶矩和三阶矩就可以了。

11.3 系统介绍

对于给定的图片或者视频流,机器人的物体检测系统可以识别出已知的物体和该物体在图片中的位置。物体检测模块被训练用于检测多种物体的存在以及它们的位置,例如模型可使用包含多个水果的图片和水果所分别代表(如苹果、香蕉、草莓)的 label 进行训练,返回的数据指明了图像中对象所出现的位置。随后，当我们为模型提供图片，模型将会返回一个列表，其中包含检测到的对象，包含对象矩形框的坐标和代表检测可信度的分数。本系统的具体结构如图 11-1 所示。

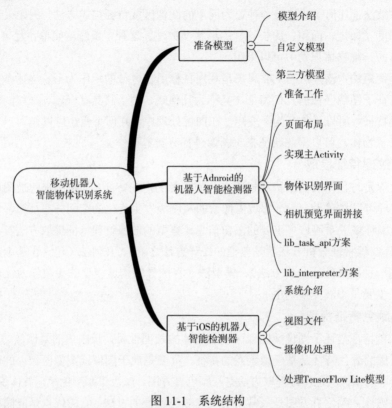

图 11-1　系统结构

11.4　准备模型

本项目使用的是 TensorFlow 官方提供的现成的模型，可以登录 TensorFlow 官方网站下载模型文件 detect.tflite。

11.4.1　模型介绍

本项目中，在文件 download_model.gradle 中设置了使用的初始模型和标签文件。文件 download_model.gradle 的具体实现代码如下：

```
task downloadModelFile(type: Download) {
    src 'https://tfhub.dev/tensorflow/lite-model/ssd_mobilenet_v1/ 1/metadata/2?lite-format=tflite'
    dest project.ext.ASSET_DIR + '/detect.tflite'
    overwrite false
}
```

物体检测模型 detect.tflite 最多能够在一张图中识别和定位 10 个物体，目前支持 80 种物体的识别。

（1）输入

模型使用单个图片作为输入，理想的图片尺寸大小是 300×300 像素，每个像素有 3 个通道（红、蓝、绿）。这将反馈给模块一个 270 000 字节（300×300×3）的扁平化缓存。由于该模块经过标准化处理，每一个字节代表了 0~255 之间的一个值。

（2）输出

该模型输出四个数组，分别对应索引的 0~4。前三个数组描述 10 个被检测到的物体，每个数组的最后一个元素匹配检测到的每个对象，检测到的物体数量总是 10。各个索引的具体说明见表 11-1。

表 11-1 索引说明

索引	名称	描述
0	坐标	[10][4]多维数组，每一个元素是 0~1 的浮点数，内部数组表示了矩形边框的[top, left, bottom, right]
1	类型	10 个整型元素组成的数组（输出为浮点型值），每一个元素代表标签文件中的索引
2	分数	10 个整型元素组成的数组，元素值为 0~1 的浮点数，代表检测到的类型
3	检测到的物体和数量	长度为 1 的数组，元素为检测到的总数

11.4.2 自定义模型

开发者可以使用转移学习等技术来重新训练模型从而能够辨识初始设置之外的物品种类，例如可以重新训练模型来辨识各种蔬菜，哪怕原始训练数据中只有一种蔬菜。为达成此目标，需要为每一个需要训练的标签准备一系列训练图片。

接下来介绍在 Oxford-IIIT Pet 数据集上训练新对象检测模型的过程，该模型将能够检测猫和狗的位置并识别每种动物的品种。本教程假设我们在 Ubuntu 16.04 系统上运行，在开始之前需要设置开发环境：

● 设置 Google Cloud 项目、配置计费并启用必要的 Cloud API。
● 设置 Google Cloud SDK。
● 安装 TensorFlow。

（1）安装 Tensorflow 对象检测 API。

假设已经安装了 Tensorflow，那么可以使用以下命令安装对象检测 API 和其他依赖项：

```
git clone https://github.com/tensorflow/models
cd models/research
sudo apt-get install protobuf-compiler python-pil python-lxml
protoc object_detection/protos/*.proto --python_out=.
export PYTHONPATH=$PYTHONPATH:`pwd`:`pwd`/slim
```

通过运行以下命令来测试安装：

```
python object_detection/builders/model_builder_test.py
```

（2）下载 Oxford-IIIT Pet Dataset。

开始下载 Oxford-IIIT Pet Dataset 数据集，然后转换为 TFRecords 并上传到 GCS。TensorFlow 对象检测 API 使用 TFRecord 格式进行训练和验证数据集。使用以下命令下载 Oxford-IIIT Pet 数据集并转换为 TFRecords：

```
wget http://www.robots.ox.ac.uk/~vgg/data/pets/data/images.tar.gz
wget http://www.robots.ox.ac.uk/~vgg/data/pets/data/annotations.tar.gz
tar -xvf annotations.tar.gz
tar -xvf images.tar.gz
python object_detection/dataset_tools/create_pet_tf_record.py \
    --label_map_path=object_detection/data/pet_label_map.pbtxt \
    --data_dir=`pwd` \
    --output_dir=`pwd`
```

接下来应该会看到两个新生成的文件：pet_train.record 和 pet_val.record。要在 GCP 上使用数据集，需要使用以下命令将其上传到我们的 Cloud Storage。注意，我们同样上传了一个"标签地图"（包含在 git 存储库中），它将我们的模型预测的数字索引与类别名称对应起来（例如，4->"basset hound", 5 -> "beagle"）。

```
gsutil cp pet_train_with_masks.record ${YOUR_GCS_BUCKET}/data/pet_train.record
gsutil cp pet_val_with_masks.record ${YOUR_GCS_BUCKET}/data/pet_val.record
gsutil cp object_detection/data/pet_label_map.pbtxt \
    ${YOUR_GCS_BUCKET}/data/pet_label_map.pbtxt
```

（3）上传用于迁移学习的预训练 COCO 模型。

从头开始训练一个物体检测器模型可能需要几天时间，为了加快训练速度，将使用提供的模型中的参数初始化宠物模型，该模型已经在 COCO 数据集上进行了预训练。这个基于 ResNet101 的 Faster R-CNN 模型的权重将成为新模型（称为微调检查点）的起点，并将训练时间从几天缩短到几个小时。要从此模型初始化，需要下载它并将其放入 Cloud Storage。

```
wget https://storage.googleapis.com/download.tensorflow.org/models/object_detection/faster_rcnn_resnet101_coco_11_06_20113.tar.gz
tar -xvf faster_rcnn_resnet101_coco_11_06_20113.tar.gz
gsutil cp faster_rcnn_resnet101_coco_11_06_2017/model.ckpt.* ${YOUR_GCS_BUCKET}/data/
```

（4）配置管道。

使用 TensorFlow 对象检测 API 中的协议缓冲区配置，可以在 object_detection/samples/configs/中找到本项目的配置文件。这些配置文件可用于调整模型和训练参数（例如学习率、dropout 和正则化参数）。需要修改提供的配置文件，以了解上传数据集的位置并微调检查点。需要更改 PATH_TO_BE_CONFIGURED 字符串，以便它们指向上传到 Cloud Storage 存储分区的数据集文件和微调检查点。之后，还需要将配置文件本身上传到 Cloud Storage。

```
sed -i "s|PATH_TO_BE_CONFIGURED|"${YOUR_GCS_BUCKET}"/data|g" object_detection/samples/configs/faster_rcnn_resnet101_pets.config
gsutil cp object_detection/samples/configs/faster_rcnn_resnet101_pets.config \
    ${YOUR_GCS_BUCKET}/data/faster_rcnn_resnet101_pets.config
```

（5）运行训练和评估。

在 GCP 上运行之前，必须先打包 TensorFlow Object Detection API 和 TF Slim。

```
python setup.py sdist
(cd slim && python setup.py sdist)
```

仔细检查是否已将数据集上传到 Cloud Storage 存储分区，可以使用 Cloud Storage 浏览器检查存储分区。目录结构应如下：

```
+ ${YOUR_GCS_BUCKET}/
  + data/
    - faster_rcnn_resnet101_pets.config
    - model.ckpt.index
    - model.ckpt.meta
    - model.ckpt.data-00000-of-00001
```

```
    - pet_label_map.pbtxt
    - pet_train.record
    - pet_val.record
```

代码打包后，准备开始训练和评估工作：

```
gcloud ml-engine jobs submit training `whoami`_object_detection_`date +%s` \
    --job-dir=${YOUR_GCS_BUCKET}/train \
    --packages dist/object_detection-0.1.tar.gz,slim/dist/slim-0.1.tar.gz \
    --module-name object_detection.train \
```

此时可以在机器学习引擎仪表板上看到作业并检查日志以确保作业正在进行中。注意，此训练作业使用具有五个工作 GPU 和三个参数服务器的分布式异步梯度下降。

（6）导出 TensorFlow 图。

为了在训练后对一些示例图像运行检测，建议尝试使用 Jupyter notebook 演示。但是，在此之前，必须将经过训练的模型导出到 TensorFlow 图形原型，并将学习到的权重作为常量进行处理。首先，需要确定要导出的候选检查点。可以使用 Google Cloud Storage Browser 搜索存储分区。检查点应存储在 "${YOUR_GCS_BUCKET} /train" 目录下。检查点通常由三个文件组成：

- model.ckpt-${CHECKPOINT_NUMBER}.data-00000-of-00001
- model.ckpt-${CHECKPOINT_NUMBER}.index
- model.ckpt-${CHECKPOINT_NUMBER}.meta

确定要导出的候选检查点（通常是最新的）后，从 "tensorflow/models" 目录运行以下命令：

```
# Please define CEHCKPOINT_NUMBER based on the checkpoint you'd like to export
export CHECKPOINT_NUMBER=${CHECKPOINT_NUMBER}

# From tensorflow/models
gsutil cp ${YOUR_GCS_BUCKET}/train/model.ckpt-${CHECKPOINT_NUMBER}.* .
python object_detection/export_inference_graph \
    --input_type image_tensor \
    --pipeline_config_path object_detection/samples/configs/faster_rcnn_resnet101_pets.config \
    --checkpoint_path model.ckpt-${CHECKPOINT_NUMBER} \
    --inference_graph_path output_inference_graph.pb
```

如果一切顺利，应该会看到导出的图形，该图形将存储在名为 output_inference_graph.pb 的文件中。

11.5 基于 Android 的机器人智能检测器

在准备好 TensorFlow Lite 模型后，接下来将使用这个模型开发一个基于 Android 系统的物体检测识别器系统。本项目提供了两种情感分析解决方案：

- lib_task_api：直接使用现成的 Task 库集成模型 API 进行 Tnference 推断识别。
- lib_interpreter：使用 TensorFlow Lite Interpreter Java API 创建自定义推断管道。

在本项目的内部 app 文件 build.gradle 中，设置了使用上述哪一种方案的方法。

11.5.1 准备工作

（1）使用 Android Studio 导入本项目源码工程 "object_detection"，如图 11-2 所示。

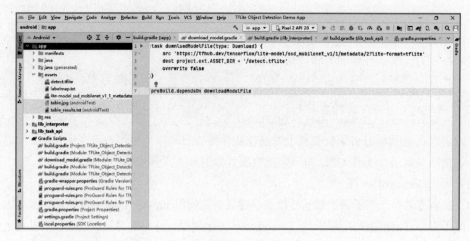

图 11-2　导入工程

（2）更新 build.gradle。

打开 app 模块中的文件 build.gradle，分别设置 Android 的编译版本和运行版本，设置需要使用的库文件，添加对 TensorFlow Lite 模型库的引用。对应的代码如下：

```
apply plugin: 'com.android.application'
apply plugin: 'de.undercouch.download'

android {
    compileSdkVersion 30
    defaultConfig {
        applicationId "org.tensorflow.lite.examples.detection"
        minSdkVersion 21
        targetSdkVersion 30
        versionCode 1
        versionName "1.0"

        testInstrumentationRunner "androidx.test.runner.AndroidJUnitRunner"
    }
    buildTypes {
        release {
            minifyEnabled false
            proguardFiles getDefaultProguardFile('proguard-android.txt'),'proguard-rules.pro'
        }
    }
    aaptOptions {
        noCompress "tflite"
    }
    compileOptions {
        sourceCompatibility = '1.8'
        targetCompatibility = '1.8'
    }
    lintOptions {
        abortOnError false
    }
    flavorDimensions "tfliteInference"
    productFlavors {
        // TFLite 推断是使用 TFLiteJava 解释器构建的
        interpreter {
```

```
            dimension "tfliteInference"
        }
        // 默认：TFLite 推断是使用 TFLite 任务库（高级 API）构建的
        taskApi {
            getIsDefault().set(true)
            dimension "tfliteInference"
        }
    }
}

//导入下载模型任务
project.ext.ASSET_DIR = projectDir.toString() + '/src/main/assets'
project.ext.TMP_DIR   = project.buildDir.toString() + '/downloads'

// 下载默认模型；如果你希望使用自己的模型，请将它们放在 "assets" 目录中，并注释掉这一行
apply from:'download_model.gradle'

dependencies {
    implementation fileTree(dir: 'libs', include: ['*.jar','*.aar'])
    interpreterImplementation project(":lib_interpreter")
    taskApiImplementation project(":lib_task_api")
    implementation 'androidx.appcompat:appcompat:1.0.0'
    implementation 'androidx.coordinatorlayout:coordinatorlayout:1.0.0'
    implementation 'com.google.android.material:material:1.0.0'
    androidTestImplementation 'androidx.test.ext:junit:1.1.1'
    androidTestImplementation 'com.google.truth:truth:1.0.1'
    androidTestImplementation 'androidx.test:runner:1.2.0'
    androidTestImplementation 'androidx.test:rules:1.1.0'
}
```

11.5.2　页面布局

（1）本项目主界面的页面布局文件是 tfe_od_activity_camera.xml，功能是在 Android 屏幕上方分别显示摄像机预览窗口，在屏幕下方显示悬浮式的系统配置参数。文件 tfe_od_activity_camera.xml 的具体实现代码如下：

```
<androidx.coordinatorlayout.widget.CoordinatorLayout xmlns:android=
"http://schemas.android.com/apk/res/android"
    xmlns:tools="http://schemas.android.com/tools"
    android:layout_width="match_parent"
    android:layout_height="match_parent"
    android:background="#00000000">

    <RelativeLayout xmlns:android="http://schemas.android.com/apk/ res/android"
        xmlns:tools="http://schemas.android.com/tools"
        android:layout_width="match_parent"
        android:layout_height="match_parent"
        android:background="@android:color/black"
        android:orientation="vertical">


        <FrameLayout xmlns:android="http://schemas.android.com/apk/ res/android"
            xmlns:tools="http://schemas.android.com/tools"
            android:id="@+id/container"
            android:layout_width="match_parent"
            android:layout_height="match_parent"
            tools:context="org.tensorflow.demo.CameraActivity" />
```

```xml
        <androidx.appcompat.widget.Toolbar
            android:id="@+id/toolbar"
            android:layout_width="match_parent"
            android:layout_height="?attr/actionBarSize"
            android:layout_alignParentTop="true"
            android:background="@color/tfe_semi_transparent">

            <ImageView
                android:layout_width="wrap_content"
                android:layout_height="wrap_content"
                android:src="@drawable/tfl2_logo" />
        </androidx.appcompat.widget.Toolbar>

    </RelativeLayout>

    <include
        android:id="@+id/bottom_sheet_layout"
        layout="@layout/tfe_od_layout_bottom_sheet" />
</androidx.coordinatorlayout.widget.CoordinatorLayout>
```

（2）在上面的页面布局文件 tfe_od_activity_camera.xml 中，通过调用文件 tfe_od_layout_bottom_sheet.xml 显示在主界面屏幕下方显示的悬浮式配置面板。文件 tfe_od_layout_bottom_sheet.xml 的主要实现代码如下：

```xml
    <LinearLayout
        android:layout_width="match_parent"
        android:layout_height="wrap_content"
        android:orientation="horizontal">

        <TextView
            android:id="@+id/frame"
            android:layout_width="wrap_content"
            android:layout_height="wrap_content"
            android:layout_marginTop="10dp"
            android:text="Frame"
            android:textColor="@android:color/black" />

        <TextView
            android:id="@+id/frame_info"
            android:layout_width="match_parent"
            android:layout_height="wrap_content"
            android:layout_marginTop="10dp"
            android:gravity="right"
            android:text="640*480"
            android:textColor="@android:color/black" />
    </LinearLayout>

    <LinearLayout
        android:layout_width="match_parent"
        android:layout_height="wrap_content"
        android:orientation="horizontal">

        <TextView
            android:id="@+id/crop"
            android:layout_width="wrap_content"
            android:layout_height="wrap_content"
            android:layout_marginTop="10dp"
            android:text="Crop"
            android:textColor="@android:color/black" />
```

```xml
        <TextView
            android:id="@+id/crop_info"
            android:layout_width="match_parent"
            android:layout_height="wrap_content"
            android:layout_marginTop="10dp"
            android:gravity="right"
            android:text="640*480"
            android:textColor="@android:color/black" />
</LinearLayout>

<LinearLayout
    android:layout_width="match_parent"
    android:layout_height="wrap_content"
    android:orientation="horizontal">

        <TextView
            android:id="@+id/inference"
            android:layout_width="wrap_content"
            android:layout_height="wrap_content"
            android:layout_marginTop="10dp"
            android:text="Inference Time"
            android:textColor="@android:color/black" />

        <TextView
            android:id="@+id/inference_info"
            android:layout_width="match_parent"
            android:layout_height="wrap_content"
            android:layout_marginTop="10dp"
            android:gravity="right"
            android:text="640*480"
            android:textColor="@android:color/black" />
</LinearLayout>

<View
    android:layout_width="match_parent"
    android:layout_height="1px"
    android:layout_marginTop="10dp"
    android:background="@android:color/darker_gray" />

<RelativeLayout
    android:layout_width="match_parent"
    android:layout_height="wrap_content"
    android:layout_marginTop="10dp"
    android:orientation="horizontal">

        <TextView
            android:layout_width="wrap_content"
            android:layout_height="wrap_content"
            android:layout_marginTop="10dp"
            android:text="Threads"
            android:textColor="@android:color/black" />

        <LinearLayout
            android:layout_width="wrap_content"
            android:layout_height="wrap_content"
            android:layout_alignParentRight="true"
            android:background="@drawable/rectangle"
            android:gravity="center"
```

```xml
            android:orientation="horizontal"
            android:padding="4dp">

            <ImageView
                android:id="@+id/minus"
                android:layout_width="wrap_content"
                android:layout_height="wrap_content"
                android:src="@drawable/ic_baseline_remove" />

            <TextView
                android:id="@+id/threads"
                android:layout_width="wrap_content"
                android:layout_height="wrap_content"
                android:layout_marginLeft="10dp"
                android:layout_marginRight="10dp"
                android:text="4"
                android:textColor="@android:color/black"
                android:textSize="14sp" />

            <ImageView
                android:id="@+id/plus"
                android:layout_width="wrap_content"
                android:layout_height="wrap_content"
                android:src="@drawable/ic_baseline_add" />
    </LinearLayout>
</RelativeLayout>
```

11.5.3　实现主 Activity

本项目的主 Activity 功能是由文件 CameraActivity.java 实现的，功能是调用前面的布局文件 tfe_od_activity_camera.xml，功能是在 Android 屏幕上方分别显示摄像机预览窗口，在屏幕下方显示悬浮式的系统配置参数。文件 CameraActivity.java 的具体实现方法如下：

（1）设置摄像头预览界面的公共属性。对应的代码如下：

```java
public abstract class CameraActivity extends AppCompatActivity
    implements OnImageAvailableListener,
        Camera.PreviewCallback,
        CompoundButton.OnCheckedChangeListener,
        View.OnClickListener {
  private static final Logger LOGGER = new Logger();

  private static final int PERMISSIONS_REQUEST = 1;

  private static final String PERMISSION_CAMERA = Manifest. permission.CAMERA;
  protected int previewWidth = 0;
  protected int previewHeight = 0;
  private boolean debug = false;
  private Handler handler;
  private HandlerThread handlerThread;
  private boolean useCamera2API;
  private boolean isProcessingFrame = false;
  private byte[][] yuvBytes = new byte[3][];
  private int[] rgbBytes = null;
```

（2）在初始化函数 onCreate() 中加载布局文件 tfe_od_activity_camera.xml。对应的代码如下：

```java
@Override
protected void onCreate(final Bundle savedInstanceState) {
```

```java
    LOGGER.d("onCreate " + this);
    super.onCreate(null);
    getWindow().addFlags(WindowManager.LayoutParams.FLAG_KEEP_SCREEN_ON);

    setContentView(R.layout.tfe_od_activity_camera);
    Toolbar toolbar = findViewById(R.id.toolbar);
    setSupportActionBar(toolbar);
    getSupportActionBar().setDisplayShowTitleEnabled(false);

    if (hasPermission()) {
      setFragment();
    } else {
      requestPermission();
    }
```

（3）获取悬浮面板中的配置参数，系统将根据这些配置参数加载显示预览界面。对应的代码如下：

```java
    threadsTextView = findViewById(R.id.threads);
    plusImageView = findViewById(R.id.plus);
    minusImageView = findViewById(R.id.minus);
    apiSwitchCompat = findViewById(R.id.api_info_switch);
    bottomSheetLayout = findViewById(R.id.bottom_sheet_layout);
    gestureLayout = findViewById(R.id.gesture_layout);
    sheetBehavior = BottomSheetBehavior.from(bottomSheetLayout);
    bottomSheetArrowImageView = findViewById(R.id.bottom_sheet_arrow);
```

（4）获取视图树观察者对象，设置底页回调处理事件。对应的代码如下：

```java
    ViewTreeObserver vto = gestureLayout.getViewTreeObserver();
    vto.addOnGlobalLayoutListener(
        new ViewTreeObserver.OnGlobalLayoutListener() {
          @Override
          public void onGlobalLayout() {
            if (Build.VERSION.SDK_INT < Build.VERSION_CODES.JELLY_BEAN) {
              gestureLayout.getViewTreeObserver(). removeGlobalOnLayoutListener(this);
            } else {
              gestureLayout.getViewTreeObserver(). removeOnGlobalLayoutListener(this);
            }
            //            int width = bottomSheetLayout.getMeasuredWidth();
            int height = gestureLayout.getMeasuredHeight();

            sheetBehavior.setPeekHeight(height);
          }
        });
    sheetBehavior.setHideable(false);

    sheetBehavior.setBottomSheetCallback(
        new BottomSheetBehavior.BottomSheetCallback() {
          @Override
          public void onStateChanged(@NonNull View bottomSheet, int newState) {
            switch (newState) {
              case BottomSheetBehavior.STATE_HIDDEN:
                break;
              case BottomSheetBehavior.STATE_EXPANDED:
                {
                  bottomSheetArrowImageView.setImageResource (R.drawable.icn_chevron_down);
                }
                break;
              case BottomSheetBehavior.STATE_COLLAPSED:
```

```java
          bottomSheetArrowImageView.setImageResource (R.drawable.icn_chevron_up);
        }
        break;
      case BottomSheetBehavior.STATE_DRAGGING:
        break;
      case BottomSheetBehavior.STATE_SETTLING:
        bottomSheetArrowImageView.setImageResource (R.drawable.icn_chevron_up);
        break;
    }
  }
  @Override
  public void onSlide(@NonNull View bottomSheet, float slideOffset) {}
});
frameValueTextView = findViewById(R.id.frame_info);
cropValueTextView = findViewById(R.id.crop_info);
inferenceTimeTextView = findViewById(R.id.inference_info);

apiSwitchCompat.setOnCheckedChangeListener(this);
plusImageView.setOnClickListener(this);
minusImageView.setOnClickListener(this);
}
```

（5）创建 android.hardware.Camera API 的回调，打开手机中的相机预览界面，使用函数 ImageUtils.convertYUV420SPToARGB8888()将相机 data 转换成 rgbBytes。对应的代码如下：

```java
@Override
public void onPreviewFrame(final byte[] bytes, final Camera camera) {
  if (isProcessingFrame) {
    LOGGER.w("Dropping frame!");
    return;
  }
  try {
    //已知分辨率，初始化存储位图一次
    if (rgbBytes == null) {
      Camera.Size previewSize = camera.getParameters(). getPreviewSize();
      previewHeight = previewSize.height;
      previewWidth = previewSize.width;
      rgbBytes = new int[previewWidth * previewHeight];
      onPreviewSizeChosen(new Size(previewSize.width, previewSize. height), 90);
    }
  } catch (final Exception e) {
    LOGGER.e(e, "Exception!");
    return;
  }
  isProcessingFrame = true;
  yuvBytes[0] = bytes;
  yRowStride = previewWidth;
  imageConverter =
      new Runnable() {
        @Override
        public void run() {
          ImageUtils.convertYUV420SPToARGB8888(bytes, previewWidth, previewHeight, rgbBytes);
        }
      };

  postInferenceCallback =
      new Runnable() {
        @Override
```

```
        public void run() {
          camera.addCallbackBuffer(bytes);
          isProcessingFrame = false;
        }
      };
   processImage();
}
```

(6) 编写函数 onImageAvailable()实现 Camera2 API 的回调。对应的代码如下：

```
@Override
public void onImageAvailable(final ImageReader reader) {
  //需要等待，直到从 onPreviewSizeChosen 得到一些尺寸
  if (previewWidth == 0 || previewHeight == 0) {
    return;
  }
  if (rgbBytes == null) {
    rgbBytes = new int[previewWidth * previewHeight];
  }
  try {
    final Image image = reader.acquireLatestImage();

    if (image == null) {
      return;
    }

    if (isProcessingFrame) {
      image.close();
      return;
    }
    isProcessingFrame = true;
    Trace.beginSection("imageAvailable");
    final Plane[] planes = image.getPlanes();
    fillBytes(planes, yuvBytes);
    yRowStride = planes[0].getRowStride();
    final int uvRowStride = planes[1].getRowStride();
    final int uvPixelStride = planes[1].getPixelStride();

    imageConverter =
        new Runnable() {
          @Override
          public void run() {
            ImageUtils.convertYUV420ToARGB8888(
                yuvBytes[0],
                yuvBytes[1],
                yuvBytes[2],
                previewWidth,
                previewHeight,
                yRowStride,
                uvRowStride,
                uvPixelStride,
                rgbBytes);
          }
        };
    postInferenceCallback =
        new Runnable() {
          @Override
          public void run() {
            image.close();
            isProcessingFrame = false;
```

```
            }
        };
        processImage();
    } catch (final Exception e) {
        LOGGER.e(e, "Exception!");
        Trace.endSection();
        return;
    }
    Trace.endSection();
}
```

（7）编写函数 onImageAvailable()，功能是判断当前手机设备是否支持所需的硬件级别或更高级别，如果是则返回 true。对应的代码如下：

```
    private boolean isHardwareLevelSupported(
        CameraCharacteristics characteristics, int requiredLevel) {
        int deviceLevel = characteristics.get(CameraCharacteristics. INFO_SUPPORTED_HARDWARE_LEVEL);
        if (deviceLevel == CameraCharacteristics.INFO_SUPPORTED_HARDWARE_ LEVEL_LEGACY) {
            return requiredLevel == deviceLevel;
        }
        //使用数字排序
        return requiredLevel <= deviceLevel;
    }
```

（8）启用当前设备中的摄像头功能。对应的代码如下：

```
    private String chooseCamera() {
        final CameraManager manager = (CameraManager) getSystemService (Context.CAMERA_SERVICE);
        try {
          for (final String cameraId : manager.getCameraIdList()) {
            final CameraCharacteristics characteristics = manager. getCameraCharacteristics(cameraId);

            //不使用前摄像头
            final Integer facing = characteristics.get(CameraCharacteristics.LENS_ FACING);
            if (facing != null && facing == CameraCharacteristics.LENS_ FACING_FRONT) {
              continue;
            }

            final StreamConfigurationMap map =
                characteristics.get(CameraCharacteristics.SCALER_STREAM_CONFIGURATION_MAP);
            if (map == null) {
              continue;
            }

            //对于没有完全支持的内部摄像头，返回 camera1 API，这将有助于解决使用 camera2 API 导致预览失真或损坏的遗留问题
            useCamera2API =
                (facing == CameraCharacteristics.LENS_FACING_EXTERNAL)
                    || isHardwareLevelSupported(
                        characteristics, CameraCharacteristics.INFO_SUPPORTED_HARDWARE_LEVEL_FULL);
            LOGGER.i("Camera API lv2?: %s", useCamera2API);
            return cameraId;
          }
        } catch (CameraAccessException e) {
          LOGGER.e(e, "Not allowed to access camera");
        }
        return null;
    }
```

11.5.4 物体识别界面

本实例的物体识别界面 Activity 是由文件 DetectorActivity.java 实现的，功能是调用 lib_task_api 或 lib_interpreter 方案实现物体识别。文件 DetectorActivity.java 的具体实现流程如下：

（1）在设置了 Camera 捕获图片的一些参数后，例如图片预览大小 previewSize，摄像头方向 sensorOrientation 等。最重要的是回调我们之前传入到 fragment 中的 cameraConnectionCallback 的 onPreviewSizeChosen()函数，这是预览图片的宽、高确定后执行的回调函数。对应的代码如下：

```java
    public void onPreviewSizeChosen(final Size size, final int rotation) {
      final float textSizePx =
          TypedValue.applyDimension(
              TypedValue.COMPLEX_UNIT_DIP, TEXT_SIZE_DIP, getResources().getDisplayMetrics());
      borderedText = new BorderedText(textSizePx);
      borderedText.setTypeface(Typeface.MONOSPACE);

      tracker = new MultiBoxTracker(this);
      int cropSize = TF_OD_API_INPUT_SIZE;
      try {
        detector =
            TFLiteObjectDetectionAPIModel.create(
                this,
                TF_OD_API_MODEL_FILE,
                TF_OD_API_LABELS_FILE,
                TF_OD_API_INPUT_SIZE,
                TF_OD_API_IS_QUANTIZED);
        cropSize = TF_OD_API_INPUT_SIZE;
      } catch (final IOException e) {
        e.printStackTrace();
        LOGGER.e(e, "Exception initializing Detector!");
        Toast toast =
            Toast.makeText(
                getApplicationContext(), "Detector could not be initialized", Toast.LENGTH_SHORT);
        toast.show();
        finish();
      }
      previewWidth = size.getWidth();
      previewHeight = size.getHeight();

      sensorOrientation = rotation - getScreenOrientation();
      LOGGER.i("Camera orientation relative to screen canvas: %d", sensorOrientation);

      LOGGER.i("Initializing at size %dx%d", previewWidth, previewHeight);
      rgbFrameBitmap = Bitmap.createBitmap(previewWidth, previewHeight, Config.ARGB_8888);
      croppedBitmap = Bitmap.createBitmap(cropSize, cropSize, Config.ARGB_8888);

      frameToCropTransform =
          ImageUtils.getTransformationMatrix(
              previewWidth, previewHeight,
              cropSize, cropSize,
              sensorOrientation, MAINTAIN_ASPECT);
```

```
cropToFrameTransform = new Matrix();
frameToCropTransform.invert(cropToFrameTransform);

trackingOverlay = (OverlayView) findViewById(R.id.tracking_ overlay);
trackingOverlay.addCallback(
    new DrawCallback() {
      @Override
      public void drawCallback(final Canvas canvas) {
        tracker.draw(canvas);
        if (isDebug()) {
          tracker.drawDebug(canvas);
        }
      }
    });
  tracker.setFrameConfiguration(previewWidth, previewHeight, sensorOrientation);
}
```

（2）处理摄像头中的图像，将流式 YUV420_888 图像转换为可理解的图，会自动启动一个处理图像的线程。这意味着可以随意使用而不会崩溃。如果你的图像处理无法跟上相机的进给速度，则会丢弃相框。对应的代码如下：

```
protected void processImage() {
    ++timestamp;
    final long currTimestamp = timestamp;
    trackingOverlay.postInvalidate();

    //不需要互斥，因为此方法不可重入
    if (computingDetection) {
      readyForNextImage();
      return;
    }
    computingDetection = true;
    LOGGER.i("Preparing image " + currTimestamp + " for detection in bg thread.");

    rgbFrameBitmap.setPixels(getRgbBytes(), 0, previewWidth, 0, 0, previewWidth, previewHeight);

    readyForNextImage();

    final Canvas canvas = new Canvas(croppedBitmap);
    canvas.drawBitmap(rgbFrameBitmap, frameToCropTransform, null);
    //用于检查实际 TF 输入
    if (SAVE_PREVIEW_BITMAP) {
      ImageUtils.saveBitmap(croppedBitmap);
    }

    runInBackground(
        new Runnable() {
          @Override
          public void run() {
            LOGGER.i("Running detection on image " + currTimestamp);
            final long startTime = SystemClock.uptimeMillis();
            final List<Detector.Recognition> results = detector. recognizeImage(croppedBitmap);
            lastProcessingTimeMs = SystemClock.uptimeMillis() - startTime;
```

```java
            cropCopyBitmap = Bitmap.createBitmap(croppedBitmap);
            final Canvas canvas = new Canvas(cropCopyBitmap);
            final Paint paint = new Paint();
            paint.setColor(Color.RED);
            paint.setStyle(Style.STROKE);
            paint.setStrokeWidth(2.0f);

            float minimumConfidence = MINIMUM_CONFIDENCE_TF_OD_API;
            switch (MODE) {
              case TF_OD_API:
                minimumConfidence = MINIMUM_CONFIDENCE_TF_OD_API;
                break;
            }

            final List<Detector.Recognition> mappedRecognitions =
                new ArrayList<Detector.Recognition>();

            for (final Detector.Recognition result : results) {
              final RectF location = result.getLocation();
              if (location != null && result.getConfidence() >= minimumConfidence) {
                canvas.drawRect(location, paint);

                cropToFrameTransform.mapRect(location);

                result.setLocation(location);
                mappedRecognitions.add(result);
              }
            }

            tracker.trackResults(mappedRecognitions, currTimestamp);
            trackingOverlay.postInvalidate();

            computingDetection = false;

            runOnUiThread(
                new Runnable() {
                  @Override
                  public void run() {
                    showFrameInfo(previewWidth + "x" + previewHeight);
                    showCropInfo(cropCopyBitmap.getWidth() + "x" + cropCopyBitmap.getHeight());
                    showInference(lastProcessingTimeMs + "ms");
                  }
                });
          }
        });
  }
```

11.5.5 摄像机预览界面拼接

编写文件 CameraConnectionFragment.java，功能是在摄像机中识别物体后会用文字标注识别结果，并将识别结果和摄像机预览界面拼接在一起，构成一幅完整的图形。文件 CameraConnectionFragment.java 的具体实现方法如下：

（1）设置长宽属性。例如，设置 MINIMUM_PREVIEW_SIZE 的属性为 320，这是为了确保相机预览大小能够容纳所需大小的正方形帧，因为 320 是最小逐像素帧的大小。对应的代码如下：

```java
    private static final int MINIMUM_PREVIEW_SIZE = 320;

    /**从屏幕旋转到 JPEG 方向的转换 */
    private static final SparseIntArray ORIENTATIONS = new SparseIntArray();

    private static final String FRAGMENT_DIALOG = "dialog";

    static {
      ORIENTATIONS.append(Surface.ROTATION_0, 90);
      ORIENTATIONS.append(Surface.ROTATION_90, 0);
      ORIENTATIONS.append(Surface.ROTATION_180, 270);
      ORIENTATIONS.append(Surface.ROTATION_270, 180);
    }

    /**一个{@link Semaphore}用于在关闭摄像头之前阻止应用程序退出*/
    private final Semaphore cameraOpenCloseLock = new Semaphore(1);
    /** 用于接收可用帧的{@link OnImageAvailableListener} */
    private final OnImageAvailableListener imageListener;
    /**TensorFlow 所需的输入大小（正方形位图的宽度和高度），以像素为单位*/
    private final Size inputSize;
    /**设置布局标识符 */
    private final int layout;
```

（2）使用 TextureView.SurfaceTextureListener 处理 TextureView 上的多个生命周期事件。对应的代码如下：

```java
    private final TextureView.SurfaceTextureListener surfaceTextureListener =
        new TextureView.SurfaceTextureListener() {
          @Override
          public void onSurfaceTextureAvailable(
              final SurfaceTexture texture, final int width, final int height) {
            openCamera(width, height);
          }

          @Override
          public void onSurfaceTextureSizeChanged(
              final SurfaceTexture texture, final int width, final int height) {
            configureTransform(width, height);
          }
          @Override
          public boolean onSurfaceTextureDestroyed(final SurfaceTexture texture) {
            return true;
          }

          @Override
          public void onSurfaceTextureUpdated(final SurfaceTexture texture) {}
        };

    private CameraConnectionFragment(
        final ConnectionCallback connectionCallback,
        final OnImageAvailableListener imageListener,
        final int layout,
        final Size inputSize) {
      this.cameraConnectionCallback = connectionCallback;
```

```
    this.imageListener = imageListener;
    this.layout = layout;
    this.inputSize = inputSize;
  }
```

（3）编写函数 chooseOptimalSize()设置摄像机的参数，会根据设置的参数返回最佳大小的预览界面。如果没有足够大的界面，则返回任意值（其中设置的宽度和高度至少与两者最小值相同）。或者如果有可能，可以选择完全匹配的值。各个参数的具体说明如下：

- choices：相机为预期输出类支持的大小列表。
- width：所需的最小宽度。
- height：所需的最小高度。

函数 chooseOptimalSize()的具体实现代码如下：

```
  protected static Size chooseOptimalSize(final Size[] choices, final int width, final int height) {
    final int minSize = Math.max(Math.min(width, height), MINIMUM_PREVIEW_SIZE);
    final Size desiredSize = new Size(width, height);

    //收集至少与预览界面一样大的支持分辨率
    boolean exactSizeFound = false;
    final List<Size> bigEnough = new ArrayList<Size>();
    final List<Size> tooSmall = new ArrayList<Size>();
    for (final Size option : choices) {
      if (option.equals(desiredSize)) {
        //设置大小，但不要返回，以便仍记录剩余的大小
        exactSizeFound = true;
      }

      if (option.getHeight() >= minSize && option.getWidth() >= minSize) {
        bigEnough.add(option);
      } else {
        tooSmall.add(option);
      }
    }

    LOGGER.i("Desired size: " + desiredSize + ", min size: " + minSize + "x" + minSize);
    LOGGER.i("Valid preview sizes: [" + TextUtils.join(", ", bigEnough) + "]");
    LOGGER.i("Rejected preview sizes: [" + TextUtils.join(", ", tooSmall) + "]");

    if (exactSizeFound) {
      LOGGER.i("Exact size match found.");
      return desiredSize;
    }

    //挑选最小的
    if (bigEnough.size() > 0) {
      final Size chosenSize = Collections.min(bigEnough, new CompareSizesByArea());
      LOGGER.i("Chosen size: " + chosenSize.getWidth() + "x" + chosenSize.getHeight());
      return chosenSize;
    } else {
      LOGGER.e("Couldn't find any suitable preview size");
      return choices[0];
    }
  }
```

(4)编写函数 showToast(),功能是显示 UI 线程上的要显示的提醒消息。对应的代码如下:

```
    private void showToast(final String text) {
      final Activity activity = getActivity();
      if (activity != null) {
        activity.runOnUiThread(
          new Runnable() {
            @Override
            public void run() {
              Toast.makeText(activity, text, Toast.LENGTH_SHORT).show();
            }
          });
      }
    }
```

(5)编写函数 setUpCameraOutputs(),功能是设置与摄像机相关的成员变量。对应的代码如下:

```
    private void setUpCameraOutputs() {
      final Activity activity = getActivity();
      final CameraManager manager = (CameraManager) activity.getSystemService(Context.CAMERA_SERVICE);
      try {
        final CameraCharacteristics characteristics = manager.getCameraCharacteristics(cameraId);

        final StreamConfigurationMap map =
            characteristics.get(CameraCharacteristics.SCALER_STREAM_CONFIGURATION_MAP);

        sensorOrientation = characteristics.get(CameraCharacteristics.SENSOR_ORIENTATION);

        //如果尝试使用过大的预览大小,可能会超过摄像机总线的带宽限制,导致华丽的预览,但会存储垃圾捕获数据
        previewSize =
            chooseOptimalSize(
                map.getOutputSizes(SurfaceTexture.class),
                inputSize.getWidth(),
                inputSize.getHeight());

        //我们将TextureView的纵横比与拾取的预览大小相匹配
        final int orientation = getResources().getConfiguration().orientation;
        if (orientation == Configuration.ORIENTATION_LANDSCAPE) {
          textureView.setAspectRatio(previewSize.getWidth(),previewSize.getHeight());
        } else {
          textureView.setAspectRatio(previewSize.getHeight(),previewSize.getWidth());
        }
      } catch (final CameraAccessException e) {
        LOGGER.e(e, "Exception!");
      } catch (final NullPointerException e) {
        //当使用Camera2API 但此代码运行的设备不支持时,会引发NPE
        ErrorDialog.newInstance(getString(R.string.tfe_od_camera_error))
            .show(getChildFragmentManager(), FRAGMENT_DIALOG);
        throw new IllegalStateException(getString(R.string.tfe_od_camera_error));
      }

      cameraConnectionCallback.onPreviewSizeChosen(previewSize,sensorOrientation);
    }
```

（6）编写函数 openCamera()，功能是打开由 CameraConnectionFragmen 指定的摄像机。对应的代码如下：

```
    private void openCamera(final int width, final int height) {
      setUpCameraOutputs();
      configureTransform(width, height);
      final Activity activity = getActivity();
      final CameraManager manager = (CameraManager) activity. getSystemService
(Context.CAMERA_SERVICE);
      try {
        if (!cameraOpenCloseLock.tryAcquire(2500, TimeUnit.MILLISECONDS)) {
          throw new RuntimeException("Time out waiting to lock camera opening.");
        }
        manager.openCamera(cameraId, stateCallback, backgroundHandler);
      } catch (final CameraAccessException e) {
        LOGGER.e(e, "Exception!");
      } catch (final InterruptedException e) {
        throw new RuntimeException("Interrupted while trying to lock camera opening.", e);
      }
    }
```

（7）编写函数 closeCamera()，功能是关闭当前的 CameraDevice 摄像机。对应的代码如下：

```
    private void closeCamera() {
      try {
        cameraOpenCloseLock.acquire();
        if (null != captureSession) {
          captureSession.close();
          captureSession = null;
        }
        if (null != cameraDevice) {
          cameraDevice.close();
          cameraDevice = null;
        }
        if (null != previewReader) {
          previewReader.close();
          previewReader = null;
        }
      } catch (final InterruptedException e) {
        throw new RuntimeException("Interrupted while trying to lock camera closing.", e);
      } finally {
        cameraOpenCloseLock.release();
      }
    }
```

（8）分别启动前台线程和后台线程。对应的代码如下：

```
  /**启动后台线程及其{@link Handler}. */
    private void startBackgroundThread() {
      backgroundThread = new HandlerThread("ImageListener");
      backgroundThread.start();
      backgroundHandler = new Handler(backgroundThread.getLooper());
    }

  /**停止后台线程及其{@link Handler}. */
    private void stopBackgroundThread() {
      backgroundThread.quitSafely();
      try {
        backgroundThread.join();
```

```
          backgroundThread = null;
          backgroundHandler = null;
        } catch (final InterruptedException e) {
          LOGGER.e(e, "Exception!");
        }
      }
```

（9）为摄像机预览界面创建新的 CameraCaptureSession 缓存。对应的代码如下：

```
    private void createCameraPreviewSession() {
      try {
        final SurfaceTexture texture = textureView.getSurfaceTexture();
        assert texture != null;

        //将默认缓冲区的大小配置为所需的相机预览大小
        texture.setDefaultBufferSize(previewSize.getWidth(),previewSize.getHeight());

        //这是我们需要开始预览的输出界面
        final Surface surface = new Surface(texture);

        //用输出曲面设置了 CaptureRequest.Builder
        previewRequestBuilder = cameraDevice.createCaptureRequest (CameraDevice.
TEMPLATE_PREVIEW);
        previewRequestBuilder.addTarget(surface);

        LOGGER.i("Opening camera preview: " + previewSize.getWidth() + "x" + previewSize.
getHeight());

        //为预览帧创建读取器
        previewReader =
            ImageReader.newInstance(
                previewSize.getWidth(), previewSize.getHeight(),ImageFormat.YUV_420_888, 2);

        previewReader.setOnImageAvailableListener(imageListener,backgroundHandler);
        previewRequestBuilder.addTarget(previewReader.getSurface());

        //为摄像机预览创建一个 CameraCaptureSession
        cameraDevice.createCaptureSession(
            Arrays.asList(surface, previewReader.getSurface()),
            new CameraCaptureSession.StateCallback() {

              @Override
              public void onConfigured(final CameraCaptureSession cameraCaptureSession) {
                //摄像机已经关闭
                if (null == cameraDevice) {
                  return;
                }

                //当会话准备就绪时开始显示预览
                captureSession = cameraCaptureSession;
                try {
                  //自动对焦，连续用于摄像机预览
                  previewRequestBuilder.set(
                      CaptureRequest.CONTROL_AF_MODE,
                      CaptureRequest.CONTROL_AF_MODE_CONTINUOUS_PICTURE);
                  // 在必要时自动启用闪存
                  previewRequestBuilder.set(
```

```
                    CaptureRequest.CONTROL_AE_MODE, CaptureRequest. CONTROL_AE_MODE_
ON_AUTO_FLASH);

                //最后，开始显示摄像机预览
                previewRequest = previewRequestBuilder.build();
                captureSession.setRepeatingRequest(
                    previewRequest, captureCallback, backgroundHandler);
              } catch (final CameraAccessException e) {
                LOGGER.e(e, "Exception!");
              }
            }
            @Override
            public void onConfigureFailed(final CameraCaptureSession cameraCapture-
Session) {
                showToast("Failed");
            }
        },
        null);
    } catch (final CameraAccessException e) {
      LOGGER.e(e, "Exception!");
    }
  }
```

（10）编写函数 configureTransform()，功能是将必要的 Matrix 转换配置为 "mTextureView"。在 setUpCameraOutputs 中确定摄像机预览大小，并且在固定 "mTextureView" 的大小后需要调用此方法。其中参数 viewWidth 表示 mTextureView 的宽度，参数 viewHeight 表示 mTextureView 的高度。对应的代码如下：

```
    private void configureTransform(final int viewWidth, final int viewHeight) {
      final Activity activity = getActivity();
      if (null == textureView || null == previewSize || null == activity) {
        return;
      }
      final int rotation = activity.getWindowManager().getDefaultDisplay().getRotation();
      final Matrix matrix = new Matrix();
      final RectF viewRect = new RectF(0, 0, viewWidth, viewHeight);
      final RectF bufferRect = new RectF(0, 0, previewSize.getHeight(), previewSize.getWidth());
      final float centerX = viewRect.centerX();
      final float centerY = viewRect.centerY();
      if (Surface.ROTATION_90 == rotation || Surface.ROTATION_270 == rotation) {
        bufferRect.offset(centerX - bufferRect.centerX(), centerY - bufferRect.centerY());
        matrix.setRectToRect(viewRect, bufferRect, Matrix.ScaleToFit. FILL);
        final float scale =
            Math.max(
                (float) viewHeight / previewSize.getHeight(),
                (float) viewWidth / previewSize.getWidth());
        matrix.postScale(scale, scale, centerX, centerY);
        matrix.postRotate(90 * (rotation - 2), centerX, centerY);
      } else if (Surface.ROTATION_180 == rotation) {
        matrix.postRotate(180, centerX, centerY);
      }
      textureView.setTransform(matrix);
    }
```

11.5.6　lib_task_api 方案

本项目默认使用 TensorFlow Lite 任务库中的开箱即用 API 实现物体检测和识别功能，通过文件 TFLiteObjectDetectionAPIModel.java 调用 Tensorflow 对象检测 API 训练的检测模型包装器，对应的代码如下：

```java
/**
 使用 Tensorflow 对象检测 API 训练的检测模型包装器
 */
public class TFLiteObjectDetectionAPIModel implements Detector {
  private static final String TAG = "TFLiteObjectDetectionAPIModelWithTaskApi";

  /** 只返回这么多结果 */
  private static final int NUM_DETECTIONS = 10;

  private final MappedByteBuffer modelBuffer;

  /**使用 Tensorflow Lite 运行模型推断的驱动程序类的实例*/
  private ObjectDetector objectDetector;

  /**用于配置 ObjectDetector 选项的生成器 */
  private final ObjectDetectorOptions.Builder optionsBuilder;

  /**
   * 初始化对图像进行分类的 TensorFlow 会话
   * {@code-labelFilename}、{@code-inputSize}和{@code-isQuantized}不是必需的，而是为了与使用 TFLite 解释器 Java API 的实现保持一致。见<a
   * *@param modelFilename 模型文件路径
   * *@param labelFilename 标签文件路径
   * *@param inputSize 图像输入的大小
   * *@param isQuantized 布尔值，表示模型是否量化
   */
  public static Detector create(
      final Context context,
      final String modelFilename,
      final String labelFilename,
      final int inputSize,
      final boolean isQuantized)
      throws IOException {
    return new TFLiteObjectDetectionAPIModel(context, modelFilename);
  }

  private TFLiteObjectDetectionAPIModel(Context context, String modelFilename) throws IOException {
      modelBuffer = FileUtil.loadMappedFile(context, modelFilename);
      optionsBuilder = ObjectDetectorOptions.builder().setMaxResults(NUM_DETECTIONS);
      objectDetector = ObjectDetector.createFromBufferAndOptions (modelBuffer, optionsBuilder.build());
  }

  @Override
  public List<Recognition> recognizeImage(final Bitmap bitmap) {
    //记录此方法，以便使用 systrace 进行分析
    Trace.beginSection("recognizeImage");
    List<Detection> results = objectDetector.detect(TensorImage.fromBitmap(bitmap));
```

```java
    // 将{@link Detection}对象列表转换为{@link Recognition}对象列表,以匹配其他推理方法的
接口,例如使用TFLite Java API
      final ArrayList<Recognition> recognitions = new ArrayList<>();
      int cnt = 0;
      for (Detection detection : results) {
        recognitions.add(
            new Recognition(
                "" + cnt++,
                detection.getCategories().get(0).getLabel(),
                detection.getCategories().get(0).getScore(),
                detection.getBoundingBox()));
      }
      Trace.endSection(); // "recognizeImage"
      return recognitions;
    }
    @Override
    public void enableStatLogging(final boolean logStats) {}

    @Override
    public String getStatString() {
      return "";
    }
    @Override
    public void close() {
      if (objectDetector != null) {
        objectDetector.close();
      }
    }

    @Override
    public void setNumThreads(int numThreads) {
      if (objectDetector != null) {
        optionsBuilder.setNumThreads(numThreads);
        recreateDetector();
      }
    }

    @Override
    public void setUseNNAPI(boolean isChecked) {
      throw new UnsupportedOperationException(
          "在此任务中不允许操作硬件加速器,只允许使用 CPU! ");
    }

    private void recreateDetector() {
      objectDetector.close();
      objectDetector    =    ObjectDetector.createFromBufferAndOptions(modelBuffer,
optionsBuilder.build());
    }
  }
```

11.5.7 lib_interpreter 方案

还可以使用 lib_interpreter 方案实现物体检测和识别功能,本方案使用 TensorFlow Lite 中的 Interpreter Java API 创建自定义识别函数。该功能主要由文件 TFLiteObjectDetection-APIModel.java 实现。对应的代码如下:

```
/**内存映射资源中的模型文件 */
```

```java
        private static MappedByteBuffer loadModelFile(AssetManager assets, String modelFilename)
            throws IOException {
        AssetFileDescriptor fileDescriptor = assets.openFd(modelFilename);
        FileInputStream inputStream = new FileInputStream(fileDescriptor.getFileDescriptor());
        FileChannel fileChannel = inputStream.getChannel();
        long startOffset = fileDescriptor.getStartOffset();
        long declaredLength = fileDescriptor.getDeclaredLength();
        return fileChannel.map(FileChannel.MapMode.READ_ONLY, startOffset, declaredLength);
    }
    /**
     * 初始化用于对图像进行分类的本机TensorFlow会话
     * *@param modelFilename 模型文件路径
     * *@param labelFilename 标签文件路径
     * *@param inputSize 图像输入的大小
     * *@param isQuantized 布尔值,表示模型是否量化
     */
    public static Detector create(
        final Context context,
        final String modelFilename,
        final String labelFilename,
        final int inputSize,
        final boolean isQuantized)
        throws IOException {
      final TFLiteObjectDetectionAPIModel d = new TFLiteObjectDetectionAPIModel();

      MappedByteBuffer modelFile = loadModelFile(context.getAssets(), modelFilename);
      MetadataExtractor metadata = new MetadataExtractor(modelFile);
      try (BufferedReader br =
          new BufferedReader(
             new InputStreamReader(
                metadata.getAssociatedFile(labelFilename),Charset.defaultCharset()))) {
        String line;
        while ((line = br.readLine()) != null) {
          Log.w(TAG, line);
          d.labels.add(line);
        }
      }

      d.inputSize = inputSize;

      try {
        Interpreter.Options options = new Interpreter.Options();
        options.setNumThreads(NUM_THREADS);
        options.setUseXNNPACK(true);
        d.tfLite = new Interpreter(modelFile, options);
        d.tfLiteModel = modelFile;
        d.tfLiteOptions = options;
      } catch (Exception e) {
        throw new RuntimeException(e);
      }

      d.isModelQuantized = isQuantized;
      //预先分配缓冲区
      int numBytesPerChannel;
      if (isQuantized) {
```

```
          numBytesPerChannel = 1; //量化
        } else {
          numBytesPerChannel = 4; //浮点数
        }
        d.imgData = ByteBuffer.allocateDirect(1 * d.inputSize * d.inputSize * 3 * numBytes-
PerChannel);
        d.imgData.order(ByteOrder.nativeOrder());
        d.intValues = new int[d.inputSize * d.inputSize];

        d.outputLocations = new float[1][NUM_DETECTIONS][4];
        d.outputClasses = new float[1][NUM_DETECTIONS];
        d.outputScores = new float[1][NUM_DETECTIONS];
        d.numDetections = new float[1];
        return d;
    }

    @Override
    public List<Recognition> recognizeImage(final Bitmap bitmap) {
      //记录此方法,以便使用 systrace 进行分析
      Trace.beginSection("recognizeImage");

      Trace.beginSection("preprocessBitmap");
      //根据提供的参数,将图像数据从0至255 int 预处理为标准化浮点
      bitmap.getPixels(intValues, 0, bitmap.getWidth(), 0, 0, bitmap.getWidth(),
bitmap.getHeight());

      imgData.rewind();
      for (int i = 0; i < inputSize; ++i) {
        for (int j = 0; j < inputSize; ++j) {
          int pixelValue = intValues[i * inputSize + j];
          if (isModelQuantized) {
            //量化模型
            imgData.put((byte) ((pixelValue >> 16) & 0xFF));
            imgData.put((byte) ((pixelValue >> 8) & 0xFF));
            imgData.put((byte) (pixelValue & 0xFF));
          } else { // Float model
            imgData.putFloat((((pixelValue >> 16) & 0xFF) - IMAGE_MEAN) / IMAGE_STD);
            imgData.putFloat((((pixelValue >> 8) & 0xFF) - IMAGE_MEAN) / IMAGE_STD);
            imgData.putFloat(((pixelValue & 0xFF) - IMAGE_MEAN) / IMAGE_STD);
          }
        }
      }
      Trace.endSection(); //预处理位图

      //将输入数据复制到 TensorFlow 中
      Trace.beginSection("feed");
      outputLocations = new float[1][NUM_DETECTIONS][4];
      outputClasses = new float[1][NUM_DETECTIONS];
      outputScores = new float[1][NUM_DETECTIONS];
      numDetections = new float[1];

      Object[] inputArray = {imgData};
      Map<Integer, Object> outputMap = new HashMap<>();
      outputMap.put(0, outputLocations);
      outputMap.put(1, outputClasses);
      outputMap.put(2, outputScores);
      outputMap.put(3, numDetections);
      Trace.endSection();
```

```java
//运行推断调用
Trace.beginSection("run");
tfLite.runForMultipleInputsOutputs(inputArray, outputMap);
Trace.endSection();

//显示最佳检测结果
//将其缩放回输入大小后,需要使用输出中的检测数,而不是顶部声明的 NUM_DETECTONS 变量
//因为在某些模型上,它们并不总是输出相同的检测总数
//例如,模型的 NUM_DETECTIONS=20,但有时它只输出 16 个预测
//如果不使用输出的 numDetections,你将获得无意义的数据
int numDetectionsOutput =
   min(
      NUM_DETECTIONS,
      (int) numDetections[0]); //从浮点转换为整数,使用最小值以确保安全

final ArrayList<Recognition> recognitions = new ArrayList<> (numDetections-
Output);
for (int i = 0; i < numDetectionsOutput; ++i) {
  final RectF detection =
     new RectF(
        outputLocations[0][i][1] * inputSize,
        outputLocations[0][i][0] * inputSize,
        outputLocations[0][i][3] * inputSize,
        outputLocations[0][i][2] * inputSize);

  recognitions.add(
     new Recognition(
        "" + i, labels.get((int) outputClasses[0][i]), outputScores[0][i],
detection));
}
Trace.endSection(); // "recognizeImage"
return recognitions;
}
```

上述两种方案的识别文件都是 Detector.java,功能是调用各自方案下面的文件 TFLiteObjectDetectionAPIModel.java 实现具体的识别功能。对应的代码如下:

```java
/**与不同识别引擎交互的通用接口 */
public interface Detector {
  List<Recognition> recognizeImage(Bitmap bitmap);
  void enableStatLogging(final boolean debug);
  String getStatString();
  void close();
  void setNumThreads(int numThreads);
  void setUseNNAPI(boolean isChecked);
  /**检测器返回的一个不变的结果,描述识别的内容 */
  public class Recognition {
    /**
     * 已识别内容的唯一标识符。特定于类,而不是对象的实例
     */
    private final String id;
    /** 用于识别的显示名称 */
    private final String title;
    /**
     * 识别度相对于其他可能性的可排序分数,分数越高越好
     */
    private final Float confidence;
    /**源图像中用于识别对象位置的可选位置*/
    private RectF location;
    public Recognition(
```

```java
        final String id, final String title, final Float confidence, final RectF
location) {
      this.id = id;
      this.title = title;
      this.confidence = confidence;
      this.location = location;
    }
    public String getId() {
      return id;
    }
    public String getTitle() {
      return title;
    }
    public Float getConfidence() {
      return confidence;
    }
    public RectF getLocation() {
      return new RectF(location);
    }
    public void setLocation(RectF location) {
      this.location = location;
    }
    @Override
    public String toString() {
      String resultString = "";
      if (id != null) {
        resultString += "[" + id + "] ";
      }
      if (title != null) {
        resultString += title + " ";
      }
      if (confidence != null) {
        resultString += String.format("(%.1f%%) ", confidence * 100.0f);
      }
      if (location != null) {
        resultString += location + " ";
      }
      return resultString.trim();
    }
  }
}
```

到此为止，整个项目工程全部开发完毕。

11.6 基于 iOS 的机器人智能检测器

在上一节讲解了基于 Android 系统为机器人开发物体检测识别器的过程，本节详细讲解在基于 iOS 系统使用 TensorFlow Lite 模型开发物体检测识别器的过程。

11.6.1 系统介绍

使用 Xcode 导入本项目的 iOS 源码，如图 11-3 所示。

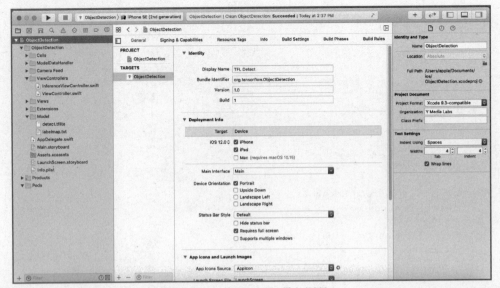

图 11-3 使用 Xcode 导入源码

在 Model 目录下保存了需要使用的 TensorFlow Lite 模型文件，如图 11-4 所示。

通过故事板 Main.storyboard 文件设计 iOS 应用程序的 UI 界面，如图 11-5 所示。

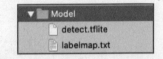

图 11-4 TensorFlow Lite 模型文件

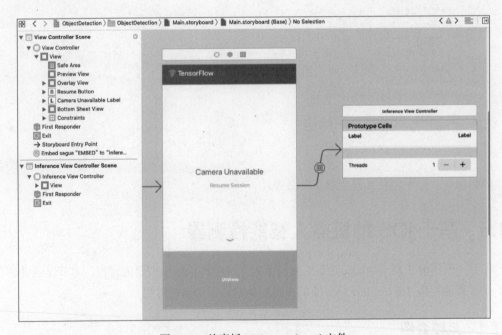

图 11-5 故事板 Main.storyboard 文件

11.6.2 视图文件

在 Xcode 工程的"ViewControllers"目录下保存了本项目的视图文件，视图文件和故事板

Main.storyboard 文件相互结合，构建 iOS 应用程序的 UI 界面。

1. 编写主视图控制器文件 ViewController.swift

具体实现方法如下：

分别设置整个系统需要的公用 UI 参数，包括连接 Storyboards 故事板中的组件参数、常量参数、实例变量和实现视图管理功能的控制器。对应的代码如下：

```swift
import UIKit

class ViewController: UIViewController {

  // MARK: 连接Storyboards故事板中的组件参数
  @IBOutlet weak var previewView: PreviewView!
  @IBOutlet weak var overlayView: OverlayView!
  @IBOutlet weak var resumeButton: UIButton!
  @IBOutlet weak var cameraUnavailableLabel: UILabel!

  @IBOutlet weak var bottomSheetStateImageView: UIImageView!
  @IBOutlet weak var bottomSheetView: UIView!
  @IBOutlet weak var bottomSheetViewBottomSpace: NSLayoutConstraint!
  // MARK: 常量
  private let displayFont = UIFont.systemFont(ofSize: 14.0, weight: .medium)
  private let edgeOffset: CGFloat = 2.0
  private let labelOffset: CGFloat = 10.0
  private let animationDuration = 0.5
  private let collapseTransitionThreshold: CGFloat = -30.0
  private let expandTransitionThreshold: CGFloat = 30.0
  private let delayBetweenInferencesMs: Double = 200
  // MARK: 实例变量
  private var initialBottomSpace: CGFloat = 0.0
  // 随时保存结果
  private var result: Result?
  private var previousInferenceTimeMs: TimeInterval = Date.distantPast.timeIntervalSince1970 * 1000
  // MARK: 管理功能的控制器
  private lazy var cameraFeedManager = CameraFeedManager(previewView: previewView)
  private var modelDataHandler: ModelDataHandler? =
    ModelDataHandler(modelFileInfo: MobileNetSSD.modelInfo, labelsFileInfo: MobileNetSSD.labelsInfo)
  private var inferenceViewController: InferenceViewController?

  //视图处理方法
  override func viewDidLoad() {
    super.viewDidLoad()

    guard modelDataHandler != nil else {
      fatalError("Failed to load model")
    }
    cameraFeedManager.delegate = self
    overlayView.clearsContextBeforeDrawing = true

    addPanGesture()
  }
  override func didReceiveMemoryWarning() {
    super.didReceiveMemoryWarning()              //处理所有可以重新创建的资源
  }
```

（1）编写函数 onClickResumeButton() 实现单击 Button 按钮后的处理程序。对应的代码如下：

```swift
@IBAction func onClickResumeButton(_ sender: Any) {
  cameraFeedManager.resumeInterruptedSession { (complete) in
    if complete {
      self.resumeButton.isHidden = true
      self.cameraUnavailableLabel.isHidden = true
    }
    else {
      self.presentUnableToResumeSessionAlert()
    }
  }
}
```

（2）编写函数 prepare() 实现故事板 Segue 处理器。对应的代码如下：

```swift
override func prepare(for segue: UIStoryboardSegue, sender: Any?) {
  super.prepare(for: segue, sender: sender)
  if segue.identifier == "EMBED" {
    guard let tempModelDataHandler = modelDataHandler else {
      return
    }
    inferenceViewController = segue.destination as? InferenceViewController
    inferenceViewController?.wantedInputHeight = tempModelDataHandler.inputHeight
    inferenceViewController?.wantedInputWidth = tempModelDataHandler.inputWidth
    inferenceViewController?.threadCountLimit = tempModelDataHandler.threadCountLimit
    inferenceViewController?.currentThreadCount = tempModelDataHandler.threadCount
    inferenceViewController?.delegate = self
    guard let tempResult = result else {
      return
    }
    inferenceViewController?.inferenceTime = tempResult.inferenceTime
  }
}
```

（3）通过 extension 扩展实现推断视图控制器。对应的代码如下：

```swift
extension ViewController: InferenceViewControllerDelegate {
  func didChangeThreadCount(to count: Int) {
    if modelDataHandler?.threadCount == count { return }
    modelDataHandler = ModelDataHandler(
      modelFileInfo: MobileNetSSD.modelInfo,
      labelsFileInfo: MobileNetSSD.labelsInfo,
      threadCount: count
    )
  }
}
```

（4）通过 extension 扩展实现摄像机管理器的委托方法。对应的代码如下：

```swift
extension ViewController: CameraFeedManagerDelegate {

  func didOutput(pixelBuffer: CVPixelBuffer) {
    runModel(onPixelBuffer: pixelBuffer)
  }
```

（5）编写自定义函数分别实现会话处理，包括实现会话处理提示框、会话中断时更新 UI、会话中断结束后更新 UI。对应的代码如下：

```swift
// MARK: 会话处理提示框
func sessionRunTimeErrorOccurred() {
```

```
      //通过更新 UI 并提供一个按钮（如果可以手动恢复会话）来处理会话运行时错误
      self.resumeButton.isHidden = false
    }
    func sessionWasInterrupted(canResumeManually resumeManually: Bool) {
      //会话中断时更新 UI
      if resumeManually {
        self.resumeButton.isHidden = false
      }
      else {
        self.cameraUnavailableLabel.isHidden = false
      }
    }
    func sessionInterruptionEnded() {
      //会话中断结束后更新 UI
      if !self.cameraUnavailableLabel.isHidden {
        self.cameraUnavailableLabel.isHidden = true
      }
      if !self.resumeButton.isHidden {
        self.resumeButton.isHidden = true
      }
    }
```

（6）如果发生错误则调用方法 presentVideoConfigurationErrorAlert()弹出提醒框。对应的代码如下：

```
    func presentVideoConfigurationErrorAlert() {
      let alertController = UIAlertController(title: "Configuration Failed", message: "Configuration of camera has failed.", preferredStyle: .alert)
      let okAction = UIAlertAction(title: "OK", style: .cancel, handler: nil)
      alertController.addAction(okAction)
      present(alertController, animated: true, completion: nil)
    }
```

（7）编写函数 presentCameraPermissionsDeniedAlert()，如果当前没有获得摄像机权限则弹出提醒框。对应的代码如下：

```
    func presentCameraPermissionsDeniedAlert() {
      let alertController = UIAlertController(title: "Camera Permissions Denied", message: "Camera permissions have been denied for this app. You can change this by going to Settings", preferredStyle: .alert)
      let cancelAction = UIAlertAction(title: "Cancel", style: .cancel, handler: nil)
      let settingsAction = UIAlertAction(title: "Settings", style: .default) { (action) in
        UIApplication.shared.open(URL(string: UIApplication.openSettingsURLString)!, options: [:], completionHandler: nil)
      }
      alertController.addAction(cancelAction)
      alertController.addAction(settingsAction)
      present(alertController, animated: true, completion: nil)
    }
```

（8）编写方法 runModel()，功能是通过 TensorFlow 运行实时摄像机像素缓冲区。对应的代码如下：

```
    @objc  func runModel(onPixelBuffer pixelBuffer: CVPixelBuffer) {

      //通过 tensorFlow 运行 pixelBuffer 摄像机以获得实时结果
      let currentTimeMs = Date().timeIntervalSince1970 * 1000
      guard (currentTimeMs - previousInferenceTimeMs) >=delayBetweenInferencesMs else {
        return
```

```
      }
      previousInferenceTimeMs = currentTimeMs
      result = self.modelDataHandler?.runModel(onFrame: pixelBuffer)
      guard let displayResult = result else {
        return
      }
      let width = CVPixelBufferGetWidth(pixelBuffer)
      let height = CVPixelBufferGetHeight(pixelBuffer)
      DispatchQueue.main.async {
        //通过传递给推断视图控制器来显示结果
        self.inferenceViewController?.resolution = CGSize(width: width, height: height)
        var inferenceTime: Double = 0
        if let resultInferenceTime = self.result?.inferenceTime {
          inferenceTime = resultInferenceTime
        }
        self.inferenceViewController?.inferenceTime = inferenceTime
        self.inferenceViewController?.tableView.reloadData()
        //绘制边界框并显示类名和置信度分数
        self.drawAfterPerformingCalculations(onInferences: displayResult.inferences, withImageSize: CGSize(width: CGFloat(width), height: CGFloat(height)))
      }
    }
```

（9）编写方法 drawAfterPerformingCalculations()获取识别结果，将边界框矩形转换为当前视图，绘制边界框、类名和推断的置信度分数。对应的代码如下：

```
    func drawAfterPerformingCalculations(onInferences inferences: [Inference], withImageSize imageSize:CGSize) {
      self.overlayView.objectOverlays = []
      self.overlayView.setNeedsDisplay()
      guard !inferences.isEmpty else {
        return
      }
      var objectOverlays: [ObjectOverlay] = []
      for inference in inferences {
        //将边界框矩形转换为当前视图
        var convertedRect = inference.rect.applying(CGAffineTransform (scaleX: self.overlayView.bounds.size.width / imageSize.width, y: self.overlayView.bounds.size.height / imageSize.height))
        if convertedRect.origin.x < 0 {
          convertedRect.origin.x = self.edgeOffset
        }
        if convertedRect.origin.y < 0 {
          convertedRect.origin.y = self.edgeOffset
        }
        if convertedRect.maxY > self.overlayView.bounds.maxY {
          convertedRect.size.height = self.overlayView.bounds.maxY - convertedRect.origin.y - self.edgeOffset
        }
        if convertedRect.maxX > self.overlayView.bounds.maxX {
          convertedRect.size.width = self.overlayView.bounds.maxX - convertedRect.origin.x - self.edgeOffset
        }
        let confidenceValue = Int(inference.confidence * 100.0)
        let string = "\(inference.className)  (\(confidenceValue)%)"
        let size = string.size(usingFont: self.displayFont)
```

```
        let objectOverlay = ObjectOverlay(name: string, borderRect: convertedRect,
nameStringSize: size, color: inference.displayColor, font: self.displayFont)
        objectOverlays.append(objectOverlay)
    }
    //将绘图交给覆盖视图
    self.draw(objectOverlays: objectOverlays)
}
```

（10）编写方法 draw()，功能是使用检测到的边界框和类名更新覆盖视图。对应的代码如下：

```
func draw(objectOverlays: [ObjectOverlay]) {
    self.overlayView.objectOverlays = objectOverlays
    self.overlayView.setNeedsDisplay()
}
```

（11）编写方法 addPanGesture()添加了平移手势处理功能，以使底部选项具有交互性。对应的代码如下：

```
private func addPanGesture() {
    let panGesture = UIPanGestureRecognizer(target: self, action: #selector
(ViewController.didPan(panGesture:)))
    bottomSheetView.addGestureRecognizer(panGesture)
}
```

（12）编写方法 changeBottomViewState()，功能是更改底部选项应处于展开还是折叠状态。对应的代码如下：

```
private func changeBottomViewState() {
    guard let inferenceVC = inferenceViewController else {
      return
    }
    if bottomSheetViewBottomSpace.constant == inferenceVC.collapsedHeight -
bottomSheetView.bounds.size.height {
      bottomSheetViewBottomSpace.constant = 0.0
    }
    else {
      bottomSheetViewBottomSpace.constant = inferenceVC.
collapsedHeight - bottomSheetView.bounds.size.height
    }
    setImageBasedOnBottomViewState()
}
```

（13）编写方法 changeBottomViewState()，功能是根据底部选项图标是展开还是折叠设置显示图像。对应的代码如下：

```
private func setImageBasedOnBottomViewState() {
    if bottomSheetViewBottomSpace.constant == 0.0 {
      bottomSheetStateImageView.image = UIImage(named: "down_icon")
    }
    else {
      bottomSheetStateImageView.image = UIImage(named: "up_icon")
    }
}
```

（14）编写方法 changeBottomViewState()响应用户在底部选项表上的平移操作。对应的代码如下：

```
@objc func didPan(panGesture: UIPanGestureRecognizer) {
    //根据用户与底部选项表的交互打开或关闭底部工作表
    let translation = panGesture.translation(in: view)
    switch panGesture.state {
```

```
    case .began:
      initialBottomSpace = bottomSheetViewBottomSpace.constant
      translateBottomSheet(withVerticalTranslation: translation.y)
    case .changed:
      translateBottomSheet(withVerticalTranslation: translation.y)
    case .cancelled:
      setBottomSheetLayout(withBottomSpace: initialBottomSpace)
    case .ended:
      translateBottomSheetAtEndOfPan(withVerticalTranslation: translation.y)
      setImageBasedOnBottomViewState()
      initialBottomSpace = 0.0
    default:
      break
    }
  }
```

（15）编写方法 translateBottomSheet() 在平移手势状态不断变化时设置底部选项平移。对应的代码如下：

```
    private func translateBottomSheet(withVerticalTranslation verticalTranslation: CGFloat) {
      let bottomSpace = initialBottomSpace - verticalTranslation
      guard bottomSpace <= 0.0 && bottomSpace >= inferenceViewController!.collapsedHeight - bottomSheetView.bounds.size.height else {
        return
      }
      setBottomSheetLayout(withBottomSpace: bottomSpace)
    }
```

（16）编写方法 translateBottomSheetAtEndOfPan()，功能是将底部选项状态更改为在平移结束时完全展开或闭合。对应的代码如下：

```
    private func translateBottomSheetAtEndOfPan(withVerticalTranslation verticalTranslation: CGFloat) {
      //将底部选项状态更改为在平移结束时完全打开或关闭
      let bottomSpace = bottomSpaceAtEndOfPan(withVerticalTranslation: verticalTranslation)
      setBottomSheetLayout(withBottomSpace: bottomSpace)
    }
```

（17）编写方法 bottomSpaceAtEndOfPan()，功能是返回要保留的底部图纸视图的最终状态（完全折叠或展开）。对应的代码如下：

```
    private func bottomSpaceAtEndOfPan(withVerticalTranslation verticalTranslation: CGFloat) -> CGFloat {
      //计算在平移手势结束时是完全展开还是折叠底部选项
      var bottomSpace = initialBottomSpace - verticalTranslation
      var height: CGFloat = 0.0
      if initialBottomSpace == 0.0 {
        height = bottomSheetView.bounds.size.height
      }
      else {
        height = inferenceViewController!.collapsedHeight
      }
      let currentHeight = bottomSheetView.bounds.size.height + bottomSpace
      if currentHeight - height <= collapseTransitionThreshold {
        bottomSpace = inferenceViewController!.collapsedHeight - bottomSheetView.bounds.size.height
      }
```

```
    else if currentHeight - height >= expandTransitionThreshold {
      bottomSpace = 0.0
    }
    else {
      bottomSpace = initialBottomSpace
    }
    return bottomSpace
}
```

（18）编写方法 setBottomSheetLayout()布局底部选项的底部空间相对于此控制器管理的视图的更改。对应的代码如下：

```
func setBottomSheetLayout(withBottomSpace bottomSpace: CGFloat) {
  view.setNeedsLayout()
  bottomSheetViewBottomSpace.constant = bottomSpace
  view.setNeedsLayout()
}
```

2. 编写推断视图控制器文件 InferenceViewController.swift

具体实现流程如下：

（1）创建继承于主视图类 UIViewController 的子类 InferenceViewController，在视图界面中显示识别信息。对应的代码如下：

```
import UIKit
protocol InferenceViewControllerDelegate {
  /**
     当用户更改步进器值以更新用于推断的线程数时，将调用此方法
   */
  func didChangeThreadCount(to count: Int)
}
class InferenceViewController: UIViewController {
  // MARK: 要显示的信息
  private enum InferenceSections: Int, CaseIterable {
    case InferenceInfo
  }
  private enum InferenceInfo: Int, CaseIterable {
    case Resolution
    case Crop
    case InferenceTime
    func displayString() -> String {
      var toReturn = ""
      switch self {
      case .Resolution:
        toReturn = "Resolution"
      case .Crop:
        toReturn = "Crop"
      case .InferenceTime:
        toReturn = "Inference Time"
      }
      return toReturn
    }
  }
  // MARK: 故事板的 Outlets 输出
  @IBOutlet weak var tableView: UITableView!
  @IBOutlet weak var threadStepper: UIStepper!
  @IBOutlet weak var stepperValueLabel: UILabel!
```

```swift
// MARK: 常量
private let normalCellHeight: CGFloat = 213.0
private let separatorCellHeight: CGFloat = 42.0
private let bottomSpacing: CGFloat = 21.0
private let minThreadCount = 1
private let bottomSheetButtonDisplayHeight: CGFloat = 60.0
private let infoTextColor = UIColor.black
private let lightTextInfoColor = UIColor(displayP3Red: 1113.0/255.0, green: 1113.0/255.0, blue: 1113.0/255.0, alpha: 1.0)
private let infoFont = UIFont.systemFont(ofSize: 14.0, weight: .regular)
private let highlightedFont = UIFont.systemFont(ofSize: 14.0, weight: .medium)

// MARK:实例变量
var inferenceTime: Double = 0
var wantedInputWidth: Int = 0
var wantedInputHeight: Int = 0
var resolution: CGSize = CGSize.zero
var threadCountLimit: Int = 0
var currentThreadCount: Int = 0
// MARK: 委托
var delegate: InferenceViewControllerDelegate?
// MARK: 计算属性
var collapsedHeight: CGFloat {
    return bottomSheetButtonDisplayHeight
}
override func viewDidLoad() {
    super.viewDidLoad()
    //设置步进器
    threadStepper.isUserInteractionEnabled = true
    threadStepper.maximumValue = Double(threadCountLimit)
    threadStepper.minimumValue = Double(minThreadCount)
    threadStepper.value = Double(currentThreadCount)
}
```

（2）将线程数的更改委托给 View Controller 并更改显示效果。对应的代码如下：

```swift
    @IBAction func onClickThreadStepper(_ sender: Any) {
        delegate?.didChangeThreadCount(to: Int(threadStepper.value))
        currentThreadCount = Int(threadStepper.value)
        stepperValueLabel.text = "\(currentThreadCount)"
    }
}
// MARK: UITableView 数据源
extension InferenceViewController: UITableViewDelegate, UITableViewDataSource {
    func numberOfSections(in tableView: UITableView) -> Int {
        return InferenceSections.allCases.count
    }
    func tableView(_ tableView: UITableView, numberOfRowsInSection section: Int) -> Int {
        guard let inferenceSection = InferenceSections(rawValue: section) else {
            return 0
        }
        var rowCount = 0
        switch inferenceSection {
        case .InferenceInfo:
            rowCount = InferenceInfo.allCases.count
        }
        return rowCount
```

```
  }
  func tableView(_ tableView: UITableView, heightForRowAt indexPath: IndexPath) -> CGFloat {
      var height: CGFloat = 0.0
      guard let inferenceSection = InferenceSections(rawValue: indexPath.section) else {
        return height
      }
      switch inferenceSection {
      case .InferenceInfo:
        if indexPath.row == InferenceInfo.allCases.count - 1 {
          height = separatorCellHeight + bottomSpacing
        }
        else {
          height = normalCellHeight
        }
      }
      return height
  }
```

(3)设置底部工作表中信息的显示格式,将格式化显示与推断相关的附加信息。对应的代码如下:

```
  func displayStringsForInferenceInfo(atRow row: Int) -> (String, String) {
      var fieldName: String = ""
      var info: String = ""
      guard let inferenceInfo = InferenceInfo(rawValue: row) else {
        return (fieldName, info)
      }
      fieldName = inferenceInfo.displayString()
      switch inferenceInfo {
      case .Resolution:
        info = "\(Int(resolution.width))x\(Int(resolution.height))"
      case .Crop:
        info = "\(wantedInputWidth)x\(wantedInputHeight)"
      case .InferenceTime:
        info = String(format: "%.2fms", inferenceTime)
      }
      return(fieldName, info)
  }
}
```

3.在 View 目录下编写文件 CurvedView.swift

功能是创建一个 CurvedView 视图,它的左上角和右上角是圆形的。具体实现代码如下:

```
import UIKit
class CurvedView: UIView {
  let cornerRadius: CGFloat = 24.0
  override func layoutSubviews() {
    super.layoutSubviews()
    setMask()
  }
  /**在视图上设置遮罩以使其拐角圆化
   */
  func setMask() {
    let maskPath = UIBezierPath(roundedRect:self.bounds,
                     byRoundingCorners: [.topLeft,.topRight],
```

```
                                    cornerRadii: CGSize(width: cornerRadius, height:
cornerRadius))
        let shape = CAShapeLayer()
        shape.path = maskPath.cgPath
        self.layer.mask = shape
    }
}
```

4. 在View目录下编写文件OverlayView.swift

功能是创建一个覆盖视图,这样可以在UI界面显示识别结果的文字内容。具体实现代码如下:

```
import UIKit
/**
此结构保存要在检测到的对象上绘制覆盖的显示参数
 */
struct ObjectOverlay {
  let name: String
  let borderRect: CGRect
  let nameStringSize: CGSize
  let color: UIColor
  let font: UIFont
}
/**
此UIView在检测到的对象上绘制覆盖
 */
class OverlayView: UIView {
  var objectOverlays: [ObjectOverlay] = []
  private let cornerRadius: CGFloat = 10.0
  private let stringBgAlpha: CGFloat = 0.7
  private let lineWidth: CGFloat = 3
  private let stringFontColor = UIColor.white
  private let stringHorizontalSpacing: CGFloat = 113.0
  private let stringVerticalSpacing: CGFloat = 13.0
  override func draw(_ rect: CGRect) {
    // 绘制代码
    for objectOverlay in objectOverlays {
      drawBorders(of: objectOverlay)
      drawBackground(of: objectOverlay)
      drawName(of: objectOverlay)
    }
  }
  /**
  此方法绘制检测到的对象的边界
   */
  func drawBorders(of objectOverlay: ObjectOverlay) {
    let path = UIBezierPath(rect: objectOverlay.borderRect)
    path.lineWidth = lineWidth
    objectOverlay.color.setStroke()
    path.stroke()
  }
  /**
  此方法绘制字符串的背景
   */
  func drawBackground(of objectOverlay: ObjectOverlay) {
    let stringBgRect = CGRect(x: objectOverlay.borderRect.origin.x, y: objectOverlay.borderRect.origin.y, width: 2 * stringHorizontalSpacing + objectOverlay.nameStringSize.
```

```
width, height: 2 * stringVerticalSpacing + objectOverlay.nameStringSize.height
    )
    let stringBgPath = UIBezierPath(rect: stringBgRect)
    objectOverlay.color.withAlphaComponent(stringBgAlpha).setFill()
    stringBgPath.fill()
  }
  /**
   此方法绘制对象覆盖的名称
   */
  func drawName(of objectOverlay: ObjectOverlay) {
    //绘制字符串
    let stringRect = CGRect(x: objectOverlay.borderRect.origin.x + stringHorizontal-
Spacing, y: objectOverlay.borderRect.origin.y + stringVerticalSpacing, width: objectOverlay.
nameStringSize.width, height: objectOverlay.nameStringSize.height)

    let attributedString = NSAttributedString(string: objectOverlay. name, attributes:
[NSAttributedString.Key.foregroundColor : stringFontColor, NSAttributedString.Key.font:
objectOverlay.font])
    attributedString.draw(in: stringRect)
  }
}
```

11.6.3 摄像机处理

在 Xcode 工程的 "Camera Feed" 目录下保存用来实现摄像机功能的程序文件，要求使用摄像机权限采集图像，然后显示采集的图像信息。

1. 编写文件 PreviewView.swift

功能是显示摄像机采集到画面的预览结果。具体实现代码如下：

```
import UIKit
import AVFoundation
/**
摄像机帧将显示在此视图上
*/
class PreviewView: UIView {
  var previewLayer: AVCaptureVideoPreviewLayer {
    guard let layer = layer as? AVCaptureVideoPreviewLayer else {
      fatalError("Layer expected is of type VideoPreviewLayer")
    }
    return layer
  }
  var session: AVCaptureSession? {
    get {
      return previewLayer.session
    }
    set {
      previewLayer.session = newValue
    }
  }
  override class var layerClass: AnyClass {
    return AVCaptureVideoPreviewLayer.self
  }
}
```

2. 编写文件 CameraFeedManager.swift 实现摄像机采集处理功能

具体实现流程如下：

（1）创建枚举保存摄像机初始化的状态。对应的代码如下：

```
enum CameraConfiguration {
  case success
  case failed
  case permissionDenied
}
```

（2）创建类 CameraFeedManager，用于管理所有与摄像机相关的功能。对应的代码如下：

```
class CameraFeedManager: NSObject {
  private let session: AVCaptureSession = AVCaptureSession()
  private let previewView: PreviewView
  private let sessionQueue = DispatchQueue(label: "sessionQueue")
  private var cameraConfiguration: CameraConfiguration = .failed
  private lazy var videoDataOutput = AVCaptureVideoDataOutput()
  private var isSessionRunning = false
  // MARK: 摄像机馈送管理器代理
  weak var delegate: CameraFeedManagerDelegate?
  // MARK: 初始化
  init(previewView: PreviewView) {
    self.previewView = previewView
    super.init()
    //初始化会话
    session.sessionPreset = .high
    self.previewView.session = session
    self.previewView.previewLayer.connection?.videoOrientation = .portrait
    self.previewView.previewLayer.videoGravity = .resizeAspectFill
    self.attemptToConfigureSession()
  }
```

（3）编写方法 checkCameraConfigurationAndStartSession()，功能是根据摄像机配置是否成功启动 AVCaptureSession。对应的代码如下：

```
func checkCameraConfigurationAndStartSession() {
  sessionQueue.async {
    switch self.cameraConfiguration {
    case .success:
      self.addObservers()
      self.startSession()
    case .failed:
      DispatchQueue.main.async {
        self.delegate?.presentVideoConfigurationErrorAlert()
      }
    case .permissionDenied:
      DispatchQueue.main.async {
        self.delegate?.presentCameraPermissionsDeniedAlert()
      }
    }
  }
}
```

（4）编写方法 stopSession()停止运行 AVCaptureSession。对应的代码如下：

```
func stopSession() {
  self.removeObservers()
  sessionQueue.async {
    if self.session.isRunning {
```

```
      self.session.stopRunning()
      self.isSessionRunning = self.session.isRunning
    }
  }
}
```

（5）编写方法 resumeInterruptedSession()恢复中断的 AVCaptureSession。对应的代码如下：
```
func resumeInterruptedSession(withCompletion completion: @escaping (Bool) -> ()) {
  sessionQueue.async {
    self.startSession()
    DispatchQueue.main.async {
      completion(self.isSessionRunning)
    }
  }
}
```

（6）编写方法 startSession()启动 AVCaptureSession。对应的代码如下：
```
private func startSession() {
  self.session.startRunning()
  self.isSessionRunning = self.session.isRunning
```

（7）编写方法 startSession()请求摄像机的权限，处理请求会话配置并存储配置结果。对应的代码如下：
```
private func attemptToConfigureSession() {
  switch AVCaptureDevice.authorizationStatus(for: .video) {
  case .authorized:
    self.cameraConfiguration = .success
  case .notDetermined:
    self.sessionQueue.suspend()
    self.requestCameraAccess(completion: { (granted) in
      self.sessionQueue.resume()
    })
  case .denied:
    self.cameraConfiguration = .permissionDenied
  default:
    break
  }
  self.sessionQueue.async {
    self.configureSession()
  }
}
```

（8）编写方法 requestCameraAccess()请求获取摄像机权限。对应的代码如下：
```
private func requestCameraAccess(completion: @escaping (Bool) -> ()) {
  AVCaptureDevice.requestAccess(for: .video) { (granted) in
    if !granted {
      self.cameraConfiguration = .permissionDenied
    }
    else {
      self.cameraConfiguration = .success
    }
    completion(granted)
  }
}
```

（9）编写方法 configureSession()处理配置 AVCaptureSession 的所有步骤。对应的代码如下：
```
private func configureSession() {
```

```
    guard cameraConfiguration == .success else {
      return
    }
    session.beginConfiguration()
    //尝试添加AVCaptureDeviceInput
    guard addVideoDeviceInput() == true else {
      self.session.commitConfiguration()
      self.cameraConfiguration = .failed
      return
    }
    //尝试添加AVCaptureVideoDataOutput
    guard addVideoDataOutput() else {
      self.session.commitConfiguration()
      self.cameraConfiguration = .failed
      return
    }
    session.commitConfiguration()
    self.cameraConfiguration = .success
  }
```

（10）编写方法 addVideoDeviceInput()，功能是尝试将 AVCaptureDeviceInput 添加到当前 AVCaptureSession。对应的代码如下：

```
    private func addVideoDeviceInput() -> Bool {
    /**尝试获取默认的后置摄像头
     */
    guard let camera = AVCaptureDevice.default
(.builtInWideAngleCamera, for: .video, position: .back) else {
      fatalError("Cannot find camera")
    }
    do {
      let videoDeviceInput = try AVCaptureDeviceInput(device: camera)
      if session.canAddInput(videoDeviceInput) {
        session.addInput(videoDeviceInput)
        return true
      }
      else {
        return false
      }
    }
    catch {
      fatalError("Cannot create video device input")
    }
  }
```

（11）编写方法 addVideoDataOutput()将 AVCaptureVideoDataOutput 添加到当前 AVCaptureSession。对应的代码如下：

```
    private func addVideoDataOutput() -> Bool {
      let sampleBufferQueue = DispatchQueue(label: "sampleBufferQueue")
      videoDataOutput.setSampleBufferDelegate(self, queue: sampleBufferQueue)
      videoDataOutput.alwaysDiscardsLateVideoFrames = true
      videoDataOutput.videoSettings = [ String(kCVPixelBufferPixelFormatTypeKey) : kCMPixelFormat_32BGRA]
      if session.canAddOutput(videoDataOutput) {
        session.addOutput(videoDataOutput)
        videoDataOutput.connection(with: .video)?.videoOrientation = .portrait
        return true
```

```
        }
        return false
    }
```

(12) 编写用于通知 Observers 观察处理器的方法 addObservers()。对应的代码如下：

```
    private func addObservers() {
        NotificationCenter.default.addObserver(self, selector: #selector(CameraFeedManager.sessionRuntimeErrorOccurred(notification:)),                                name: NSNotification.Name.AVCaptureSessionRuntimeError, object: session)
        NotificationCenter.default.addObserver(self, selector: #selector(CameraFeedManager.sessionWasInterrupted(notification:)),                                name: NSNotification.Name.AVCaptureSessionWasInterrupted, object: session)
        NotificationCenter.default.addObserver(self, selector: #selector(CameraFeedManager.sessionInterruptionEnded), name: NSNotification.Name.AVCaptureSessionInterruptionEnded, object: session)
    }
    private func removeObservers() {
        NotificationCenter.default.removeObserver(self, name: NSNotification.Name.AVCaptureSessionRuntimeError, object: session)
        NotificationCenter.default.removeObserver(self, name: NSNotification.Name.AVCaptureSessionWasInterrupted, object: session)
        NotificationCenter.default.removeObserver(self, name: NSNotification.Name.AVCaptureSessionInterruptionEnded, object: session)
    }
    // MARK: 通知 Observers
    @objc func sessionWasInterrupted(notification: Notification) {
        if let userInfoValue = notification.userInfo?[AVCaptureSessionInterruptionReasonKey] as AnyObject?,
            let reasonIntegerValue = userInfoValue.integerValue,
            let reason = AVCaptureSession.InterruptionReason(rawValue: reasonIntegerValue) {
            print("Capture session was interrupted with reason \(reason)")
            var canResumeManually = false
            if reason == .videoDeviceInUseByAnotherClient {
              canResumeManually = true
            } else if reason== .videoDeviceNotAvailableWithMultipleForegroundApps {
              canResumeManually = false
            }
            self.delegate?.sessionWasInterrupted(canResumeManually: canResumeManually)
        }
    }

    @objc func sessionInterruptionEnded(notification: Notification) {
      self.delegate?.sessionInterruptionEnded()
    }

    @objc func sessionRuntimeErrorOccurred(notification: Notification) {
        guard let error = notification.userInfo?[AVCaptureSessionErrorKey] as? AVError else {
            return
        }
        print("Capture session runtime error: \(error)")
        if error.code == .mediaServicesWereReset {
            sessionQueue.async {
                if self.isSessionRunning {
                    self.startSession()
```

```
        } else {
          DispatchQueue.main.async {
            self.delegate?.sessionRunTimeErrorOccurred()
          }
        }
      }
    } else {
      self.delegate?.sessionRunTimeErrorOccurred()
    }
  }
}
```

(13)创建扩展 CameraFeedManager,功能是将 AVCapture 视频数据输出样本缓冲区委托,通过 captureOutput()方法输出摄像机当前看到的帧的 CVPixelBuffer。对应的代码如下:

```
extension CameraFeedManager: AVCaptureVideoDataOutputSampleBufferDelegate {
  func captureOutput(_ output: AVCaptureOutput, didOutput sampleBuffer: CMSampleBuffer,
from connection: AVCaptureConnection) {

    //将CMSampleBuffer 转换为CVPixelBuffer
    let pixelBuffer: CVPixelBuffer? = CMSampleBufferGetImageBuffer (sampleBuffer)

    guard let imagePixelBuffer = pixelBuffer else {
      return
    }
    //将像素缓冲区委托给ViewController
    delegate?.didOutput(pixelBuffer: imagePixelBuffer)
  }
}
```

11.6.4 处理 TensorFlow Lite 模型

在 Xcode 工程的"ModelDataHandler"目录下编写文件 ModelDataHandler.swift,用于使用 TensorFlow Lite 模型实现物体检测识别功能。具体实现流程如下:

(1)定义结构体 Result,存储通过"Interpreter"实现成功物体识别的结果。对应的代码如下:

```
struct Result {
  let inferenceTime: Double
  let inferences: [Inference]
}
```

(2)使用 Inference 存储一个格式化的推断。对应的代码如下:

```
struct Inference {
  let confidence: Float
  let className: String
  let rect: CGRect
  let displayColor: UIColor
}
///有关模型文件或标签文件的信息
typealias FileInfo = (name: String, extension: String)
```

(3)通过枚举 MobileNetSSD 存储有关 MobileNet SSD 型号的信息。对应的代码如下:

```
enum MobileNetSSD {
  static let modelInfo: FileInfo = (name: "detect", extension: "tflite")
  static let labelsInfo: FileInfo = (name: "labelmap", extension: "txt")
}
```

（4）定义类 ModelDataHandler 处理所有的预处理数据，并通过调用"Interpreter"在给定帧上运行推断。然后格式化获得的推断结果，并返回成功推断中的前 N 个结果。对应的代码如下：

```swift
class ModelDataHandler: NSObject {
  // MARK: -内部属性
///TensorFlow Lite解释器使用的当前线程计数
  let threadCount: Int
  let threadCountLimit = 10
  let threshold: Float = 0.5
  // MARK: 模型参数
  let batchSize = 1
  let inputChannels = 3
  let inputWidth = 300
  let inputHeight = 300
  //浮动模型的图像平均值和标准差应与模型训练中使用的参数一致
  let imageMean: Float = 1213.5
  let imageStd:  Float = 1213.5

  // MARK: 私有属性
  private var labels: [String] = []

  /// TensorFlow Lite 的"Interpreter"对象，用于对给定模型执行推理
  private var interpreter: Interpreter

  private let bgraPixel = (channels: 4, alphaComponent: 3, lastBgrComponent: 2)
  private let rgbPixelChannels = 3
  private let colorStrideValue = 10
  private let colors = [
    UIColor.red,
    UIColor(displayP3Red: 90.0/255.0, green: 200.0/255.0, blue: 250.0/255.0, alpha: 1.0),
    UIColor.green,
    UIColor.orange,
    UIColor.blue,
    UIColor.purple,
    UIColor.magenta,
    UIColor.yellow,
    UIColor.cyan,
    UIColor.brown
  ]
```

（5）编写方法 init?()实现初始化操作，设置"ModelDataHandler"的可失败初始值设定项。如果从应用程序的主捆绑包成功加载模型和标签文件，则会创建一个新实例。默认的 threadCount 值为 1。对应的代码如下：

```swift
    init?(modelFileInfo: FileInfo, labelsFileInfo: FileInfo, threadCount: Int = 1) {
      let modelFilename = modelFileInfo.name
      //构造模型文件的路径
      guard let modelPath = Bundle.main.path(
        forResource: modelFilename,
        ofType: modelFileInfo.extension
      ) else {
        print("Failed to load the model file with name: \(modelFilename).")
        return nil
      }
      //指定该选项的`Interpreter`选项
      self.threadCount = threadCount
```

```swift
      var options = Interpreter.Options()
      options.threadCount = threadCount
      do {
        // 创建`Interpreter`
        interpreter = try Interpreter(modelPath: modelPath, options: options)
        //为模型输入'Tensor'的分配内存
        try interpreter.allocateTensors()
      } catch let error {
        print("Failed to create the interpreter with error: \(error.localizedDescription)")
        return nil
      }
      super.init()
      //加载标签文件中列出的类
      loadLabels(fileInfo: labelsFileInfo)
    }
```

（6）编写方法 runModel()处理所有的预处理数据，并通过 Interpreter 调用在指定的帧上运行推断。然后，格式化处理推断结果，并返回成功推断中的前 N 个结果。对应的代码如下：

```swift
    func runModel(onFrame pixelBuffer: CVPixelBuffer) -> Result? {
      let imageWidth = CVPixelBufferGetWidth(pixelBuffer)
      let imageHeight = CVPixelBufferGetHeight(pixelBuffer)
      let sourcePixelFormat = CVPixelBufferGetPixelFormatType(pixelBuffer)
      assert(sourcePixelFormat == kCVPixelFormatType_32ARGB ||
             sourcePixelFormat == kCVPixelFormatType_32BGRA ||
             sourcePixelFormat == kCVPixelFormatType_32RGBA)
      let imageChannels = 4
      assert(imageChannels >= inputChannels)

      //将图像裁剪到中心最大的正方形，并将其缩小到模型尺寸
      let scaledSize = CGSize(width: inputWidth, height: inputHeight)
      guard let scaledPixelBuffer = pixelBuffer.resized(to: scaledSize) else {
        return nil
      }
      let interval: TimeInterval
      let outputBoundingBox: Tensor
      let outputClasses: Tensor
      let outputScores: Tensor
      let outputCount: Tensor
      do {
        let inputTensor = try interpreter.input(at: 0)
        //从图像缓冲区中删除 alpha 组件以获取 RGB 数据
        guard let rgbData = rgbDataFromBuffer(
          scaledPixelBuffer,
          byteCount: batchSize * inputWidth * inputHeight * inputChannels,
          isModelQuantized: inputTensor.dataType == .uInt8
        ) else {
          print("Failed to convert the image buffer to RGB data.")
          return nil
        }
        //将 RGB 数据复制到输入张量
        try interpreter.copy(rgbData, toInputAt: 0)
        //调用`Interpreter`.
        let startDate = Date()
        try interpreter.invoke()
```

```swift
        interval = Date().timeIntervalSince(startDate) * 1000
        outputBoundingBox = try interpreter.output(at: 0)
        outputClasses = try interpreter.output(at: 1)
        outputScores = try interpreter.output(at: 2)
        outputCount = try interpreter.output(at: 3)
    } catch let error {
        print("Failed to invoke the interpreter with error: \(error.localizedDescription)")
        return nil
    }
    //格式化结果
    let resultArray = formatResults(
        boundingBox: [Float](unsafeData: outputBoundingBox.data) ?? [],
        outputClasses: [Float](unsafeData: outputClasses.data) ?? [],
        outputScores: [Float](unsafeData: outputScores.data) ?? [],
        outputCount: Int(([Float](unsafeData: outputCount.data) ?? [0])[0]),
        width: CGFloat(imageWidth),
        height: CGFloat(imageHeight)
    )
    //返回推断时间和推断
    let result = Result(inferenceTime: interval, inferences: resultArray)
    return result
}
```

（7）编写方法 formatResults() 筛选出置信度"得分<阈值"的所有结果，返回按降序排序的前 N 个结果。对应的代码如下：

```swift
func formatResults(boundingBox: [Float], outputClasses: [Float], outputScores: [Float], outputCount: Int, width: CGFloat, height: CGFloat) -> [Inference]{
    var resultsArray: [Inference] = []
    if (outputCount == 0) {
        return resultsArray
    }
    for i in 0...outputCount - 1 {

        let score = outputScores[i]

        //筛选 confidence < threshold 的结果
        guard score >= threshold else {
            continue
        }
        //从标签列表中获取检测到的类的输出类名
        let outputClassIndex = Int(outputClasses[i])
        let outputClass = labels[outputClassIndex + 1]
        var rect: CGRect = CGRect.zero
        //将检测到的边界框转换为 CGRect
        rect.origin.y = CGFloat(boundingBox[4*i])
        rect.origin.x = CGFloat(boundingBox[4*i+1])
        rect.size.height = CGFloat(boundingBox[4*i+2]) - rect.origin.y
        rect.size.width = CGFloat(boundingBox[4*i+3]) - rect.origin.x
        //检测到的角点用于模型尺寸，所以我们根据实际的图像尺寸来缩放 rect
        let newRect = rect.applying(CGAffineTransform(scaleX: width, y: height))
        //获取为类指定的颜色
        let colorToAssign = colorForClass(withIndex: outputClassIndex + 1)
        let inference = Inference(confidence: score,
                                  className: outputClass,
```

```
                       rect: newRect,
                       displayColor: colorToAssign)
    resultsArray.append(inference)
}
//排序结果按可信度的降序排列
resultsArray.sort { (first, second) -> Bool in
    return first.confidence > second.confidence
}
return resultsArray
```

(8) 编写方法 loadLabels()加载标签, 并将其存储在 "labels" 属性中。对应的代码如下:

```
private func loadLabels(fileInfo: FileInfo) {
    let filename = fileInfo.name
    let fileExtension = fileInfo.extension
    guard let fileURL = Bundle.main.url(forResource: filename, withExtension: fileExtension) else {
        fatalError("Labels file not found in bundle. Please add a labels file with name " +
            "\(filename).\(fileExtension) and try again.")
    }
    do {
        let contents = try String(contentsOf: fileURL, encoding: .utf8)
        labels = contents.components(separatedBy: .newlines)
    } catch {
        fatalError("Labels file named \(filename).\(fileExtension) cannot be read. Please add a " +
            "valid labels file and try again.")
    }
}
```

(9) 编写方法 rgbDataFromBuffer()返回具有指定值的给定图像缓冲区的 RGB 数据表示形式, 各个参数的说明如下:

- buffer: 用于转换为 RGB 数据的 BGRA 像素缓冲区。
- byteCount: 使用模型的训练内容为: 'batchSize*imageWidth*imageHeight* ComponentScont'。
- isModelQuantized: 模型是否量化（即固定点值而非浮点值）。

返回值是图像缓冲区的 RGB 数据表示形式, 如果无法创建缓冲区则返回 ni。方法 rgbDataFromBuffer()的实现代码如下:

```
private func rgbDataFromBuffer(
    _ buffer: CVPixelBuffer,
    byteCount: Int,
    isModelQuantized: Bool
) -> Data? {
    CVPixelBufferLockBaseAddress(buffer, .readOnly)
    defer {
        CVPixelBufferUnlockBaseAddress(buffer, .readOnly)
    }
    guard let sourceData = CVPixelBufferGetBaseAddress(buffer) else {
        return nil
    }
    let width = CVPixelBufferGetWidth(buffer)
    let height = CVPixelBufferGetHeight(buffer)
    let sourceBytesPerRow = CVPixelBufferGetBytesPerRow(buffer)
    let destinationChannelCount = 3
```

```
        let destinationBytesPerRow = destinationChannelCount * width
        var sourceBuffer = vImage_Buffer(data: sourceData,
                            height: vImagePixelCount(height),
                            width: vImagePixelCount(width),
                            rowBytes: sourceBytesPerRow)
        guard let destinationData = malloc(height * destinationBytesPerRow) else {
          print("Error: out of memory")
          return nil
        }
        defer {
          free(destinationData)
        }
        var destinationBuffer = vImage_Buffer(data: destinationData,
                            height: vImagePixelCount(height),
                            width: vImagePixelCount(width),
                            rowBytes: destinationBytesPerRow)
        if (CVPixelBufferGetPixelFormatType(buffer) == kCVPixelFormatType_32BGRA){
          vImageConvert_BGRA8888toRGB888(&sourceBuffer, &destinationBuffer, UInt32(kvImageNoFlags))
        } else if (CVPixelBufferGetPixelFormatType(buffer) == kCVPixelFormatType_32ARGB) {
          vImageConvert_ARGB8888toRGB888(&sourceBuffer, &destinationBuffer, UInt32(kvImageNoFlags))
        }
        let byteData = Data(bytes: destinationBuffer.data, count: destinationBuffer.rowBytes * height)
        if isModelQuantized {
          return byteData
        }
        //未量化,转换为浮点数
        let bytes = Array<UInt8>(unsafeData: byteData)!
        var floats = [Float]()
        for i in 0..<bytes.count {
          floats.append((Float(bytes[i]) - imageMean) / imageStd)
        }
        return Data(copyingBufferOf: floats)
      }
```

(10) 编写方法 colorForClass() 为特定类指定颜色。对应的代码如下:

```
      private func colorForClass(withIndex index: Int) -> UIColor {
        //有一组颜色,会根据每个对象的索引为各个对象指定基础颜色的变化
        let baseColor = colors[index % colors.count]
        var colorToAssign = baseColor
        let percentage = CGFloat((colorStrideValue / 2 - index / colors.count) * colorStrideValue)
        if let modifiedColor = baseColor.getModified(byPercentage: percentage) {
          colorToAssign = modifiedColor
        }
        return colorToAssign
      }
```

(11) 创建扩展 Data,功能是给指定数组的缓冲区指针创建新缓冲区。对应的代码如下:

```
      extension Data {
        init<T>(copyingBufferOf array: [T]) {
          self = array.withUnsafeBufferPointer(Data.init)
        }
      }
```

（12）创建扩展 Array，功能是根据指定不安全数据的字节创建新的数组。对应的代码如下：

```
extension Array {
  init?(unsafeData: Data) {
    guard unsafeData.count % MemoryLayout<Element>.stride == 0 else { return nil }
    #if swift(>=5.0)
    self = unsafeData.withUnsafeBytes { .init($0.bindMemory(to: Element.self)) }
    #else
    self = unsafeData.withUnsafeBytes {
      .init(UnsafeBufferPointer<Element>(
        start: $0,
        count: unsafeData.count / MemoryLayout<Element>.stride
      ))
    }
    #endif // swift(>=5.0)
  }
}
```

另外，在 Xcode 工程的"Cells"目录下编写文件 InfoCell.swift，功能是使用单元格形式显示识别结果列表。具体实现代码如下：

```
import UIKit
class InfoCell: UITableViewCell {
  @IBOutlet weak var fieldNameLabel: UILabel!
  @IBOutlet weak var infoLabel: UILabel!
}
```

11.7 调试运行

无论是在 Android 机器人设备，还是在 iOS 机器人设备，运行后都可以实时显示自带摄像机中物体的识别结果，执行效果如图 11-6 所示。

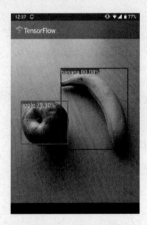

图 11-6 执行效果